Copyright (C) 2000

Printed in the United States of America

ISBN 1-58545-070-7

We are interested in hearing from authors with book ideas.

Published by The Nafziger Collection, Inc.
PO Box 1522, West Chester, OH 45069-1522

E-mail: Nafziger@fuse.net
On-line Catalog: http://home.fuse.net/nafziger

About this volume

A great deal of thought was given to the manner in which to present this material. There were two options. Onc was to read and consolidate it into a digested review of the original documents and the other was to simply present the source documents in their original form with editorial commentaries only when absolutely necessary for a clear understanding of the topic at hand or when the current intelligence was so inaccurate as to require correction. The latter was chosen because that preserves the flavor and to allow individuals seeking to use this material for their own research to maintain traceability to the original source documents.

The first work stands by itself and was not broken up and consolidated into the topical divisions as were the Intelligence Bulletins. However, the Intelligence Bulletins generally covered so many diverse topics that it was found preferable to segregate the specific tactical topics by subject and present them in a more coherent manner. For the sake of reference traceability, however, the page numbers of the original document are indicated at the beginning of each topic.

TANK TALK
GERMAN TANK TRENDS

Just what can be expected from German tanks in the near future? Which models are most likely to be employed extensively?   Are present models undergoing much alternation?

A brief summary of the German tank situation at the moment should serve to answer these and other pertinent questions.

There is good reason to believe that the German tanks which will be most frequently encountered in the near future will be the Pz. Kpfw. V (Panther), and the Pz. Kpfw. VI (Tiger) and the Pz.Kpfw. IV. However, the Germans have a new 88mm (3.46 inch) tank gun, the KwK 43, which is capable of an armor piercing performance superior to that of the 88mm KwK 36. According to reliable information, the KwK 43, which is superseding the KwK 36 as the main armament of the Tiger. A new heavy tank, which has been encountered on a small scale in northwestern France, also is armed with the KwK 43. This new tank looks like a scaled-up Panther, with the wide Tiger tracks. (Further information regarding this tank will appear in an early issue of the *Intelligence Bulletin*).

During recent months both the Tiger and Panther have been fitted with a slightly more powerful 690-horsepower engine in place of the 642-horsepower model. The principal benefit from this slight increase will be a better margin of power and improved engine line. The maximum speed will be increased by no more than 2 or 3 miles per hour.

Face hardened armor, which was not used on the early Tiger tanks, has reappeared in certain plates of at least one Panther. On other Panthers which have been encountered, only machine quality armor is used. There is no reason to believe that face-hardening would substantially improve armor's resistance to penetration by the capped projectiles now in use against it.

It would not have been surprising if the Pz. Kpfw. IV had slowly disappeared from the picture as increased quantities of Panther tanks became available, but actually there was a sharp rise in the rate of production of Pz. Kpfw. IV's during 1943. Moreover, the front armor of the Pz. Kpfw. IV has been reinforced from 50mm (1.97 inches) to 80mm (3.15 inches) by the bolting of additional armor to the nose and front vertical plates. And the 75mm (2.95 inch) tank gun, KwK 40, has been lengthened by about 14¾-inches.

All these developments seem to indicate the Pz. Kpfw IV probably will be kept in service for many months. Recent organization evidence reflects this, certainly. In the autumn of 1943, evidence regarding provisional organization for the German tank regiment in an armored division indicated that the aim was a ration of approximately four Panther tanks for each Pz Kpfw. IV. How, however, the standard tank regiment has these two types in approximately equal numbers.

The possibility that Tiger production may have been discontinued has been considered. Although discontinuing the Tiger would relieve the pressure on German industry, it is believed that a sufficient number of these tanks to meet the needs of units equipped with them still is being produced.

Tiger tanks constitute an integral part of division tank regiments only in SS armored divisions. However, armored divisions of an army may receive an allotment of Tigers for special operations.

Early in 1944 a number of Pz. Kpfw. III's converted into flame-throwing tanks appeared in Italy. Nevertheless, it is believed that production of this tank ceased some time ago. Some of the firms which in the past produced Pz. Kpfw. III's are now making assault guns; others are believed to be turning out Panthers. It is extremely unlikely that production of Pz. Kpfw. III's as fighting tanks will ever be resumed, no matter how serious the German tank situation may become.

In an effort to combat attacks by tank hunters, the Germans have fitted the Tiger with S-mine dischargers, which are fired electronically from the interior of the tank. These dischargers are mounted on the turret, and are designed to provide a shrapnel antipersonnel mine which bursts in the air a few yards away from the tanks. Thus far these dischargers have been noted only on the Tiger, but the Germans quite possibly may decide to use them on still other tanks.

The Germans take additional precautions, as well. For protection against hollow-charge projectiles and the Soviet antitank rifle's armor-piercing bullet with a tungsten carbide core, they fit a skirting of mild steel plates, about ¼-inch thick, on the sides of the hull. In the case of the Pz. Kpfw. IV, the skirting is suitably spaced from the sides and also from the rear of the turret. Finally, the skirting plates, as well as the hulls and turrets of the tanks, themselves, are coated with a sufficient thickness of non-magnetic plaster to prevent magnetic demolition charges from adhering to the metal underneath.

Despite the recent introduction of the new heavy tank which resembles the Panther and mounts a KwK 43, it is believed that circumstances will force the Germans to concentrate on the manufacture and improvement of current types, particularly the Pz. Kpfw. IV and the familiar version of the Panther.

Evidence suggests that a modified Pz.Kpfw. II will shortly appear as a reconnaissance vehicle. Official German documents sometimes refer to it as an armored car and sometimes as a tank.

GERMAN TANKS IN ACTION

A German prisoner observes that the following are standard training principals in the German tank arm:

1. Surprise.
2. Prompt decisions and prompt execution of these decisions.
3. The fullest possible exploitation of the terrain for firing. However, fields of fire come before cover.
4. Do not fire while moving except when absolutely essential.
5. Face the attacker head-on; do not offer a broadside target.
6. When attacked by hostile tanks, concentrate solely on these.
7. If surprised without hope of favorable defense, scatter and reassemble in favorable terrain. Try to draw the attacker into a position which will give you an advantage.

8. If smoke is to be used, keep wind direction in mind. A good procedure is to leave a few tanks in position as decoys, and, when the hostile force is approaching them, to direct a smoke screen toward the hostile force to blind it.
9. If hostile tanks are sighted, German tanks should halt and prepare to engage them by surprise, holding fire as long as possible. The reaction of the hostile force must be estimated before the attack is launched.

A German document entitled, "How the Tiger Can Aid the Infantry" contains a number of interesting points. The following are outstanding:

1. The tank expert must have a chance to submit his opinion before any combined tank-infantry attack.
2. If the ground will support a man standing on one leg and carrying another man on his shoulders, it will support a tank.
3. When mud is very deep, corduroy roads must be built ahead of time. Since this requires manpower, material, and time, the work should be undertaken only near the point where the main effort is to be made.
4. Tanks must be deployed to conduct their fire fight.
5. The Tiger, built to fight tanks and antitank guns, must function as an offensive weapon, even in the defense. This is its best means of defense against hostile tanks. Give it a chance to use its unique capabilities for fire and movement.
6. The Tiger must keep moving. At the halt it is an easy target.
7. The Tiger must not be used singly. [Obviously, this does not apply to the Tiger used as roving artillery in the defense. On numerous occasions the Germans have been using single Tigers for this purpose.] The more mass you can assemble, the greater your success will be. Protect your Tigers with infantry.

HOW TO FIGHT PANZERS: A GERMAN VIEW

An anti-Nazi prisoner of war, discussing the various methods of combating German tanks, makes some useful comments. Although they are neither new nor startling, they are well worth studying since they are observations made by a tank man who fought the United Nations forces in Italy.

German tanks undoubtedly are formidable weapons against a softshelled opposition, but become a less difficult proposition when confronted with resolution combined with a knowledge not only of their potentialities but also of their weaknesses.

When dealing with German heavy tanks, your most effective weapon is your ability to keep still and wait for them to come within effective range. The next most important thing is to camouflage your position with the best available resource so that the German tanks won't spot you from any angle.

If these two factors are constantly kept in mind, the battle is half won. Movement of any kind is a mistake which certainly will betray you, yet I saw many instances of this self-betrayal by the British in Italy. Allow the enemy tank to approach as close as possible before engaging it -- this is one of the fundamental secrets of antitank success. In Italy I often felt that the British opened fire on tanks much too soon. Their aim was good, but the ranges were too great, and the rounds failed to penetrate. My own case is a good illustration: If the opposition had held its fire for only a few moments longer, I should not have been alive to tell this tale.

By letting the German tank approach as close as possible, you gain a big advantage. When it is on the move, it is bound to betray its presence from afar, whereas you yourself can prepare to fire on it without giving your position away. The tank will spot you only after you have fired your first round.

A tank in motion cannot fire effectively with its cannon; the gunner can place fire accurately only when the vehicle is stationary. Therefore, there is no need to be unduly nervous because an approaching tank swivels its turret this way and that. Every tank commander will do this in an attempt to upset his opponents' tank recognition. If the tank fires nothing but its machine guns, you can be pretty sure that you have not yet been spotted.

Consider the advantages of firing on a tank at close range:

1. In most cases the leading tank is a reconnaissance vehicle. Survivors of the crew, when such a short distance away from you, have little chance of escape. This is a big advantage, inasmuch as they cannot rejoin their outfit and describe the location of your position to the main body.

2. Another tank following its leader on a road cannot run you down. In order to bypass the leading tank, it has to slow down. Then, long before the gunner can place fire on you, you can destroy the tank and block the road efficiently. Earlier in the war, a German tank man I knew destroyed 11 hostile tanks in one day by using this method.

VULNERABILITY OF THE PZ.KPFW. IV

A tank is such a complicated weapon, with its many movable parts and its elaborate mechanism, that it is particularly valuable to know its points of greatest vulnerability. Recently the *Soviet Artillery Journal* published a number of practical suggestions, based on extensive combat experience, regarding the vulnerability of the Tiger.

All weapons now used for destroying German tanks -- antitank guns and rifles, caliber .50 heavy machine guns, antitank grenades, and Molotov cocktails -- are effective against the Pz. Kpfw. IV.

1. *Suspension System.* -- The mobility of tanks depends upon the proper functioning of the suspension parts: the sprocket (small driving wheel), the idler (small wheel in the rear), the wheels, and the tracks. All these parts are vulnerable to shells of all calibers. The sprocket is especially vulnerable.

Fire armor-piercing shells and high explosives at the sprocket, idler, and tracks.

Fire at the wheels with high-explosive shells. Use antitank grenades, antitank mines, and movable antitank mines against the suspension parts. Attach three or four mines to a board. Place the board wherever tanks are expected to pass. Camouflage the board in the proper direction and place it under the track of the tank.

[A German source states that this method was successfully used on roads and road crossings in Russia, and that it is still taught in tank combat courses for infantry. The mine is called the *Scharniermine* (pivot mine). It consists of a stout length of board, 8 inches wide by 2 inches thick, and cut to a length dependent on the width of the road to be blocked. A hole is bored at one end, through which a spike or bayonet can be driven into the ground, thus providing a pivot for the board. A hook is fastened to the other end of the board, and a rope is tied to the hook, as shown Figure 3. Tellermines are secured to the top of the board.

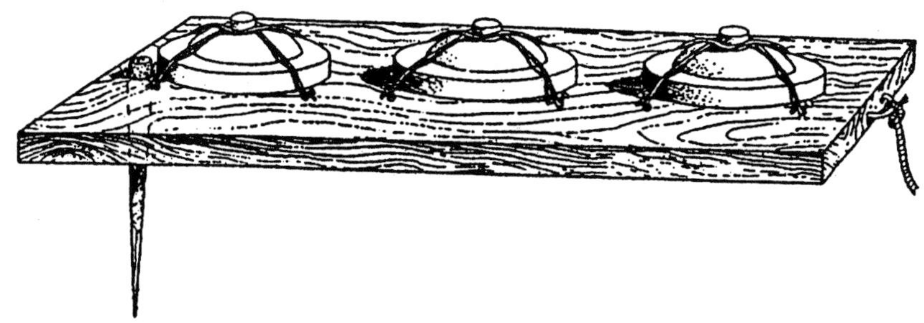

Figure 3.

One man can operate this mine. After the board has been fastened down at one end with the spike (in emergencies, a bayonet) and a rope tied to the hook at the other end, the board is laid along the side of the road. On the opposite side of the road, a man is posted in a narrow slit trench. He holds the other end of the rope. When a tank approaches, the tank hunter waits until it is close enough to the pivoted board, and, at the very last moment, he pulls the free end of the board across the road. The rope and slit trench must be well camouflaged. A good emphasis is placed on this point.]

2. *Side Armor Plates.* -- There are two armor plates on each side of the tank. The lower plate is partially covered by the wheels. This plate protects the engine and the gasoline tanks, which are located in the rear of the hull -- directly beyond and over the two rear wheels. Ammunition is kept in special compartments along the side of the tank. The compartments are protected by the upper armor plate.

Fire armor-piercing shells from 76m, 57mm and 45mm guns at the upper and lower armor plate. When the gas tanks or ammunition compartments are hit, the vehicle will be set on fire.

3. *Rear Armor Plate.* -- The rear armor plate protects the engine, the gasoline tanks, and the radiators.

Use antitank guns. Aim at the rear armor plate. When the engine or the gasoline tanks are hit, the tank will halt and will begin to burn.

4. *Peepholes, Vision Ports, and Slits.* -- The main turret has two openings for firing small-arms weapons, and two vision ports. The smaller turret has five observation slits. There are two sighting devices on the roof of the front of the tank -- one for the driver, the other for the gunner. There is also a port with sliding covers in the front armor plate.

Use all available weapons for firing at the peepholes, observation ports, vision slits, and the ports for small-arms weapons.

5. *Turrets.* -- The commander's turret is an important and vulnerable target.

Fire high-explosive and armor-piercing shells of all calibers at the commander's turret. Throw antitank grenades and incendiary bottles after the turret has been damaged.

The tank commander, the turret commander, and the gunner ride in the turret. The tank gun and many mechanical devices are found in the turret.

Fire at the turret with 76mm, 57mm, and 45mm shells at ranges of 500 yards or less.

6. *Tank Armament.* -- The turret is armed with a gun and a machine gun mounted co-axially. Another machine gun is found in the front part of the hull. It protrudes through the front armor plate, on a ball mount, and is manned by the radio operator.

Concentrate the fire of all weapons on the armament of the tank. Fire with antitank rifles at the ball mount of the hull machine gun.

7. *Air Vents and Ventilators.* -- The air vents and ventilators are found under the slit-shaped perforations in the roof of the hull, directly behind the turret. Another air vent is located in the front part of the roof, between the two observation ports used by the radio operator and the driver.

Use incendiary bottles and antitank grenades to damage the ventilating system.

8. *Tank Floor.* -- When an antitank mine explodes under the tank, the floor of the tank is smashed, and the tank is knocked out of action.

9. *Base of Turret.* -- There is a 10mm slit going all around the turret, between the base of the turret and the roof of the hull.

Fire at the base of the turret with heavy machine guns and antitank guns, to destroy the turret mechanism and disrupt the field of fire. Fire with high-explosive shells at the base of the turret in order to wreck the roof of the hull and put the tank out of action.

Intelligence Bulletin Vol II, No 10 (June 1944), pages 24-28

Section VII. GERMAN TANK PLATOON OPERATING AS POINTS

This section discusses the composition and employment of German tank platoons operating as points. Although the information in this account comes from an unofficial source, it is believed to be substantially correct.

1. COMPOSITION

The point platoon is generally made up of the platoon leader's tank and two sections of two tanks each. The platoon leader may place either the first or second section at the head of the point platoon, but he himself always stays between the two sections in order to observe his entire outfit. However, the composition of the point varies according to the situation.

The strength of the point platoon may be increased in mountainous terrain. During the German invasion of the Balkans, the point amounted to an extra-strong company and consisted of heavy tanks, assault weapons, tanks with long 75mm and 50mm guns, an infantry platoon, and a detachment of engineers. A platoon of five Pz.Kw. IV's led the point. Behind them came a group of engineers, riding either on the last tanks in the point or on other tanks immediately following. After that came a platoon of self-propelled assault guns (four short-barreled 75mm's), then the platoon of infantry riding in armored personnel carriers, and finally a platoon of five Pz.Kw. III's. There were no motorcycle couriers.

At the historic Thermopylae Pass, in Greece, there were 22 tanks in the spearhead, but only three of these got through. A responsible German's officer's comment was that it was worth losing the 19 tanks in order to achieve success with the three.

2. COMMUNICATION

a. Within the Point Platoon

In combat, communication within the German tank platoon operating as a point is done basically by radio. Up to that time, liaison is maintained by at least one or two motorcycle couriers attached to the platoon leader. As soon as contact with a hostile force is established, these couriers scatter to the sides and lie in ditches until the whole company has passed. They then go back to the company commander and report to him that contact has been made. After this, he carries on by radio.

b. Within the Armored Regiment

As has been stated, there are five tanks in each platoon -- two in each section and one for the platoon leader. The platoon leader and each section leader has a two-way radio;' the two remaining tanks have receiving sets only. Regimental commanders and all three battalion commanders have special radio cars, each equipped with 100-watt sets. If the battalions (or companies) attack together, they have radio communication with the regiment. When they attack separately, each uses, in addition to his two-way radio (*Funk Gerät* 5), four sets capable only of receiving (*Funk Gerät* 2's). Each of these receiving sets is issued for communication with one of the four companies. Moreover, each company is on a different frequency. In turn each company commander has a two-way set and two receiving sets, and can speak with the battalion commander.

Each battalion, too, is normally on a different frequency. The platoon is on the same frequency as its company commander. Each platoon leader has his second receiving set tuned to the frequency of his battalion commander, in case his company commander should become a casualty.

If the regiment attacks as a unit, the network remains unchanged. However, if the battalions act independently, the regimental commander has no communication with them except by messengers, usually motorcyclists.

Code is used only with the 100-watt sets, from battalion up to division. During the attack, communication is in the clear, even up to the regimental commander. When battalions attack separately, however, they use code in communicating with the regimental commander.

The division commander alone authorizes messages in the clear. If the battalion commander cannot reach his regimental commander by using the two-way *Funk Gerät* 6 (which has a range of 6 kilometers), he encodes his message and uses the 100-watt set.

3. ON THE MARCH

a. Combat Vehicles

It is a German principal that the distance between the rear of the point platoon and the company commander must not be so great that the latter cannot see the former. It can be, but seldom is, as much as 1 kilometer. The spacing depends entirely on the terrain. All movement is made by road until a hostile force is encountered. The tanks then scatter to the sides. Even when there is danger of air attack, the tanks remain on the road, but keep well apart. In mountainous country, when heavy tanks are used in the point, the method of advancing on roads is altered. Two tanks advance together, one behind the other but on the opposite side of the road.

The sections are easily interchangeable; for example, should the first section be at the head of the platoon and then leave the road to overcome hostile resistance, the second section can move to the head, allowing the first section to fall in behind when the resistance has been overcome. The Germans believe that it is of the utmost importance to keep the platoon moving forward.

b. Supply Column

During the campaign in Greece, all supply trucks were placed in the rear. In any other position they would have delayed movement, because of the twisting mountainous roads. Any truck that was damaged was immediately shoved off the road to keep the column moving at all costs.

In more recent operations, when facing the possibility of a guerrilla attack from the front (rather than from the flank), the Germans have been known to sandwich elements of the supply column between tank platoons on the march. The important ration and fuel trucks have even traveled between tanks within a platoon. While this plan has not been followed by a point platoon, it has been employed by platoons following immediately afterward in the line of march. The same plan has occasionally been used by German battalions on the march, but only when there has been a danger of attacks by guerrillas or when road conditions have been so bad that supply trucks have needed tanks close at hand at all times for emergency towing.

Intelligence Bulletin Vol II, No 4 (December 1943), pages 57-63

Section IV. USE OF TANKS WITH INFANTRY

1. INTRODUCTION

The correct and incorrect ways of using infantry with tanks, according to the German Army view, are summarized in an enemy document recently acquired. In this document the Germans list the correct and incorrect methods side by side, an arrangement which is also followed in this section, for the convenience of the reader. The document is of special value and interest, not only because the column headed "Right" indicates procedures approved by the enemy, but because there are implications, in the column headed "Wrong," of certain errors that German units may have made from time to time.

Extracts from the document follow.

2. THE DOCUMENT

a. Attack	*Right*
Wrong Attack not thoroughly discussed in advance.	1. Thorough discussions of reconnaissance and terrain will take place. Riflemen and tanks will maneuver jointly as much as possible, in advance.
Inadequate coordination between armored and artillery units.	2. The mission of protecting armored elements not yet discovered by hostile forces will be distributed among artillery. Flanks will be screened by smoke.
Failure of armored cars and tanks to maneuver jointly in advance.	3. Armored cars used for observation will maneuver with tanks before an intended attack.
Distribution of too many tanks in proportion to infantry used in the attack.	4. Tanks not intended for use in an attack will be kept outside the range of hostile fire.
Tanks deployed and distributed among small units.	5. For effective results, available tanks – at least an entire company – will be combined for the assault.
The use of tanks in unreconnoitered terrain when speed is essential.	6. Terrain must be reconnoitered, especially when an attack at great speed is contemplated. Facilities for mine clearance must be at hand. If a tank detonates a mine, the remaining tanks must halt while the minefield is reconnoitered. After this, the minefield must either be cleared or bypassed.
All tank commanders absent on reconnaissance.	7. A number of tank commanders must always remain with the company.
Tanks launched without a clear statement of their mission.	8. The mission of tanks will be widely understood.
When a sector full of tank obstacles has been taken, tanks are ordered to cross this sector in front of the riflemen.	9. Riflemen cross the sector first and create passages, while the tanks provide covering fire from positions on slopes.

Tanks advance so rapidly that riflemen are unable to follow.

When two successive objectives have been taken, tanks ignore possible presence of hostile forces in areas between these objectives, even though an attack on still another objective is not contemplated at the moment.

Tanks within sight of positioned hostile tanks advance without benefit of covering fire.

Tanks are ordered to hold a captured position, even though heavy weapons are available for this purpose.

Riflemen and light machine guns remain under cover during own attack.

Tanks take up positions so close to hostile forces that early discovery is inevitable

Tanks remain inactive when a mission has been completed.

b. Defense
Distribution of tanks along the entire front.

---

10. Tanks advance only a short distance at a time. Riflemen advance with the tanks.
11. When two successive objectives have been taken, the entire area between them must be made secure by means of tanks, artillery, assault guns, or antitank guns, and heavy weapons.
12. Responding for covering fire is divided among artillery or heavy antitank guns. If these are not available, PzKw. 3's and Pz.Kw. 4's provide protection
13. As soon as an objective has been taken, tanks are withdrawn and kept in readiness for use as an attacking reserve or in the preparation of a new attack.
14. Riflemen and machine guns cover the antitank riflemen, who have the mission of destroying hostile tanks which may attempt to bypass.
15. If possible, tanks take up positions outside the range of hostile artillery fire. Tanks which are compelled to take up positions in the vicinity of hostile forces do so as late as possible, so that the hostile forces will not have time to adopt effective countermeasures.
16. When a mission has been completed, tanks promptly receive orders as to what they are to do next.

1. All available tanks are kept together so that during an enemy attack prompt action can be taken against an advantageous point. Tanks, assault guns, and heavy antitank guns must be kept at a distance while firing positions are being prepared.

Subordination of tanks to small infantry units for the purposes of static defense.

After repulsing an attack, tanks remain in the positions from which they last fired.

As hostile tanks approach, own tanks advance, having failed to take up advantageous firing positions beforehand.

Tanks which have no armor-piercing weapons are sent into battle against hostile tanks.

When hostile tanks approach, German riflemen and their heavy arms remain under cover, and leave the fighting against tanks with infantry to own tanks, assault guns, and antitank guns exclusively.

All available tank reserves are compelled to remain out of action because of minor defects.
Tanks which must remain in forward positions do not dig in, and thereby constitute targets for hostile artillery.

c. Notes on Use of Ammunition

When only a few hostile tanks attack, fire is opened early.

Against a superior number of tanks, fire is opened at close range.

2. When tanks have fulfilled their task they are withdrawn behind the main line of resistance and are kept in readiness for further action.
3. After repulsing an attack, tanks move to alternative positions as soon as heavy arms or riflemen have taken over the responsibility of delivering covering fire.
4. A firing front is created at a tactically advantageous point in the area against which the attack is directed. Tanks deliver surprise fire – from positions on reverse slopes, if possible.
5. Tanks without armor-piercing weapons are kept back, and are used for antiaircraft protection, as well as establishing communications and in supplying ammunition.
6. All arms take part in defense against hostile tanks. Infantry accompanying the tanks are kept somewhat apart, however, so that tanks, assault guns, and antitank guns are free to engage the hostile tanks.
7. Repairs will be arranged in such a manner that a number of tanks are always ready for action.
8. Tanks which are within range of hostile observation must be dug in as fast as possible. In winter, they must be hidden behind snow walls.

1. When only a few enemy tanks attack, it is best to wait until they are within a favorable distance and then destroy them with as few rounds as possible.
2. Fire is opened early on a superior number of tanks, to force them to change direction. High-explosive shells are used at first. Since this early opening of fire gives away

Pz.Kw 4's will fire hollow charge ammunition at ranges of more than 750 yards.	own positions, new positions must be taken up. 3. Tanks which are short of 75mm armor-piercing shells must allow a hostile force to approach a position within a range of 750 yards.
d. Peculiarities of Winter Fighting	
Tanks are placed outside "tank shelters" when the shelters are being used for other purposes. In deep snow, tanks do not advance on roads.	1. "Tank shelters" are to be kept for the exclusive use of tanks, assault guns, and mounted anti-tank guns. 2. In deep snow, tanks keep to roads. An adequate number of men are detailed to assist if fresh snow falls.
Winter quarters are located so far from the scene of action that the tanks, if required, may arrive too late.	3. When action in appreciably distant places is under consideration, arrangements must be made for the smaller units – if possible, never less than a platoon – to reach the scene of action at the proper time.
When "tank shelters" are snowed under, departure is possible only after hours of extra labor.	4. Paths leading from "tank shelters" to the nearest roads are kept cleared. Snow fences are provided for exits. Readiness of tanks is always assured.
In winter, tanks travel freely over roads which have not been used for a considerable time.	5. Because of danger from land mines, mine-clearance detachments always precede tanks, especially if a road is seldom used.
In winter, tanks are ordered to attack distant objectives.	6. All attacks consist of a number of consecutive attacks with "limited objectives." When these objectives have been reached, the area is cleared and reorganization is completed before a new attack is launched.

Intelligence Bulletin Vol II, No 60 (February 1944), pages 75-77

Section V. NOTES ON GERMAN ANTITANK TACTICS

1. ANTITANK METHODS IN RUSSIA

The following observations represent an authoritative Soviet view of German antitank methods.

The German antitank defenses open up while our [Soviet] armor is moving toward the front line or when it has reached its line of departure. First, German bombers and artillery go into action to halt our attack, or at least to delay it.

The German artillery (GHQ units, divisional units, and in rare instances regimental guns) lays down a barrage about 2 miles inside our lines, and tries to smash our armor. Each German battery is assigned a frontage of about 100 to 150 yards, which it must cover. When our tanks are within 200 to 300 yards of the antitank obstacles on our side of the German main defensive area, the German guns transfer their fire to the accompanying Soviet infantry.

When our tanks are within 600 to 1,000 yards of the German main defensive area, single antitank guns (chiefly regimental) are brought into action. The main antitank strength opens up only when the range has been reduced still further, and is between 300 and 150 yards. The guns which constitute the main strength are sited principally for enfilade fire from well-camouflaged positions.

The Germans site most of their antitank weapons to the rear of the forward edge of their main defensive area. Only single guns are sited along the forward edge; their mission is to engage individual tanks. As soon as an attack has been repelled, these guns change position. Antitank reserves are placed in areas most vulnerable to tank attack, especially at boundaries between units. Infantry antitank reserves consist of a platoon of antitank guns and several tank-hunting detachments, and are sometimes reinforced by infantry, field guns and tanks.

Positions are planned for all-around defense. Two or three alternate positions are prepared for each antitank gun. Roving guns are used extensively, especially in the less vital areas. Assault guns and self-propelled antitank guns are used, not only as a mobile antitank reserve, but also as fixed weapons dug-in near the forward edge of the main defensive zone.

The main antitank weapon strength is concentrated against the flanks and rear of the attacking tanks. Gun positions are protected by antitank mines and by tank-hunting detachments. Very often, too, the Germans mine the ruts made by retreating tanks, in the hope that Soviet tanks will use them as a guide.

As the Soviet tanks reach the German main defensive line, tank -hunting detachments go into action. At this stage smoke may be used, but only if the antitank guns have ceased firing, inasmuch as smoke hinders accurate laying. When the tanks reach the German gun positions, the field guns fire over open sights.

2. ENGAGING TANKS AT CLOSE RANGE

The following order was issued by the general officer commanding the Fifteenth Panzer Division during the last days of the Tunisia fighting:

"The general officer commanding the Army Group Africa desires that, as a rule, the antitank artillery engage hostile armored vehicles at ranges of not more than 600 yards, and that special attention be paid to close-range engagement of tanks by tank-hunting detachments. I repeat my instruction that training in close-range engagement of tanks with all weapons shall be stressed. Every man in this division who knocks out a tank in close combat will receive the Assault Badge and, in addition, a special leave."

3. AN ANTITANK COMPANY LAYOUT

The following description of a German antitank company layout was provided by a prisoner of war. Since this layout would be dictated entirely by terrain factors, it should be regarded as an instance of enemy flexibility, rather than as a typical arrangement.

"Platoons were in line, with their guns echeloned. Each platoon had tow guns forward, about 200 yards apart, and a third gun to the rear, equidistant from the other two. The distance to the nearest gun of the adjoining platoon was about 300 yards. On each side of the gun position, there was a light machine gun, in line with the forward antitank guns and about 30 yards from the nearest neighboring gun."

Intelligence Bulletin Vol III, No 6. (January 1945), pages 10-18

MORE NOTES ON ANTITANK TACTICS

A German two-man bazooka team firing on U.S. Tanks

With German antitank activity coming increasingly into the spotlight, these new notes on enemy antitank tactics have a special significance. Moreover, it must be expected that such measures will become even more vigorous as the threat to the German homeland grows.

U.S. combat experiences in Italy and interrogation of German officers have yielded fresh information about German antitank tactics, which are playing a more important part than ever in the enemy's stubborn defensive fighting. The following tactical notes deal with the antitank company, the bazooka and grenade-discharger squads, and ground-mounted antitank guns, tanks, and self-propelled artillery in the withdrawal. In addition, a new German technique of preparing delaying positions is discussed and illustrated. The latter information comes from a U.S. armored division now fighting in Germany.

THE ANTITANK COMPANY

Companies of the German division antitank battalion, as well as the regimental 14th Company, are employed in support of the infantry regiments, but their orders for deployment normally come from the antitank battalion headquarters, rather than from the regiment. The Germans believe that this procedure ensures a higher degree of coordination in the antitank defense throughout the division sector. However, the following tactical principles are followed by companies of both types.

The guns are brought into an assembly area, and the company and platoon commanders go forward to make a detailed reconnaissance of firing positions. If the company commander has had enough time, he will have prepared a map designating areas as *Panzerschier* (tank-proof), *Panzergefährdet* (difficult for tanks), or *Panzermöglich* (good tank terrain). The over-all allotment of antitank guns will have been made on the basis of this map, with the object of covering those areas designated as *Panzermöglich*. Great care is taken with the siting of each gun; whenever possible, this is done by the platoon commander.

The caliber of the gun determines the nature of the positions which are chosen. The Germans stipulate that the 50mm [Pak 38] antitank gun must be sited in defilade and must fire to the flanks. This is why the Germans choose such positions as reverse slopes of hills and the reverse edges of small woods. Houses are avoided, on the principal that they attract too much artillery fire. The Germans also prefer flanking fire for their 75mm [Pak 40] antitank guns, but the U.S. and British practice of advancing with infantry in the lead and tanks following in support makes this difficult to achieve. And since the Germans believe that these 75's can pierce the front armor of Allied tanks at ranges up to 2,000 yards, the guns usually are sited to fire forward, and are well camouflaged instead of being defiladed. Guns of all calibers are sited in depth, at varying distances behind the main line of resistance, depending on the situation and the terrain. Invariably, the guns have infantry in front of them for local protection. The positions are arranged so that the guns can support each other, each gun covering positions from which other guns might be attacked by Allied tanks in hull-down positions.

The enemy's normal practice has been to withdraw the company's prime movers to lines about half a mile to a mile behind the gun positions, but, because of Allied heavy artillery superiority, this is no longer possible. The more usual procedure is to send back all but one prime mover out of range of artillery fire. The remaining prime mover serves for any local changes of position which prove necessary. Since such changes of position are likely to be fairly frequent, it is standard enemy practice to prepare alternate positions for the guns as soon as the original positions have been prepared. Of course, a company with only one prime mover forward is not able to undertake a sudden withdrawal. (If the probable necessity for a withdrawal is foreseen, the prime movers are kept near the guns, and are disposed in whatever cover can be found.)

If the company makes a planned withdrawal, assembly areas as well as new lines of resistance are reconnoitered to the rear of the initial position. The guns then withdraw singly, under cover of remaining weapons.

BAZOOKA AND GRENADE-DISCHARGER SQUADS

Bazooka and grenade-discharger squads are allotted to those infantry companies whose sectors are considered most likely to be attacked by tanks. The bazooka is regarded as a relatively static weapon, to be fired from a prepared position, whereas the grenade discharger is regarded as a mobile reserve weapon and usually is held back at the Antitank Company command post.

Wherever possible, bazookas are used in groups of three, and are sited in a V with its prongs toward the opposition.

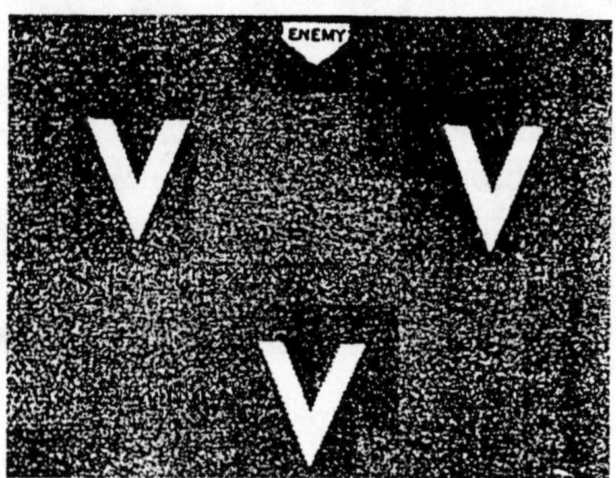

This permits at least two weapons to engage a tank approaching from any direction. The individual rocket launcher is emplaced in a V-shaped pit, with the prongs pointing toward the opposition. The weapon is carried to either end of the V, according to the direction from which the tank to be engaged is approaching. Having loaded the weapon, the loader takes shelter in the opposite arm of the V, to avoid the back-flash of the rocket. He usually is armed with a machine gun, and is responsible for the ground protection of the position. In the general defense plan of the company, bazooka sections have the mission of defending narrow tank lanes and defiladed approaches. The antitank guns cover the open areas of attack.

A "FIRE TEAM" IN THE WITHDRAWAL

Ground-mount antitank guns, tanks, and self-propelled artillery frequently constitute a "fire team" in the German withdrawal actions. The ground-mount antitank guns are sited singly, in groups of two or three, in positions permitting all-around defense. Wire entanglements and minefields surround these positions, and infantry in company or platoon strength is maintained in the immediate vicinity. The infantry stays close to the road, and their positions are planned for ready withdrawal. For this reason the infantry engage mostly with frontal fire, and fire only a few rounds before moving back. Small groups of tanks deploy on the flanks of their position, serving both as protection for the antitank guns and as an incentive to hostile armor to deploy similarly. In retreat, these tanks engage the hostile armor and afford time for the ground-mount weapons to retire to their next position.

The mission of the self-propelled guns in an action of this kind is to remain in the rear, between the antitank guns in the center and the armor on the flanks. The self-propelled guns provide fire support, changing position continually and avoiding a direct engagement with the hostile armor. The German consider them especially valuable in helping antitank guns to defend a road block. By changing their positions so often, the self-propelled guns place interdictory fire of heavy caliber on the obstacle area without endangering themselves to any appreciable extent.

FAILURE OF A TACTIC

A prisoner declared that the antitank company never was used in support of advancing tanks; its chief mission was to attack Allied tanks and cover the German retreat.

The three guns of a platoon are staggered in the following manner.

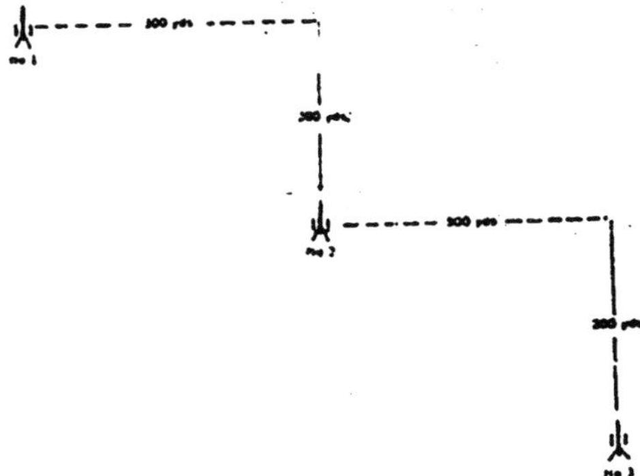

In theory, the No. 1 gun was to start firing so as to attract return fire from the hostile tanks. As soon as it was feasible to do so, the No. 1 gun was to change position to the rear. In the meantime, the No. 2 gun was to fire in order to attract the opposition's attention. As soon as the Allied tanks' fire was directed toward the No. 2 gun, that gun was supposed to cease firing and move to the rear, leaving the No. 3 gun to take over until the No. 1 gun was in position and ready to start firing again.

In actual practice, however, this system seldom worked. Fire on advancing tanks was opened at 400 yards; the prisoner considered this range much too short to permit a successful change of position as outlined in the theory. Since the prisoner was captured by advancing Allied infantry because he had been unable to move his gun to the rear quickly enough, his contention seems pretty reasonable.

No spare barrels were carried by the enemy, and only the gun sight was used. The prisoner's platoon once had a range finder calibrated up to 10,000 yards, but the prisoner had never seen it in action. Fire control was independent for each gun, and was handled by the noncom in charge, who relied on field glasses.

In the prisoner's opinion, the following ranges for the 75mm [Pak 40] antitank guns were the most effective:

Against tanks or other moving targets	400 yards
Against attacking infantry	1,000 yards
Against strong points	1,200 yards
Against houses	1,500 yards

A DELAYING POSITION

In recent weeks a U.S. armored division has been encountering German delaying positions designed to destroy the leading tanks of an armored column and to cause confusion and delay. One type of set-up in particular has been encountered repeatedly, and evidence shows that the Germans have been practicing and perfecting the technique very studiously indeed.

As shown in the illustration, a covered and perfectly camouflaged foxhole for a two-man bazooka team normally is dug in a semicircular pattern around a corner of a house or other building, anywhere from 5 to 50 yards off a road. A camouflaged escape trench leads from the village of the bazooka emplacement to any nearby place of concealment, such as garden shrubbery, outbuildings, or woods. Machine guns are placed in a V, with the prongs of the V about 300 to 400 yards away from the road and facing the direction from which an Allied approach is expected.

When an advancing Allied column is preceded by a dismounted point, firing is withheld until the bazooka team can be certain of knocking out the leading tanks.

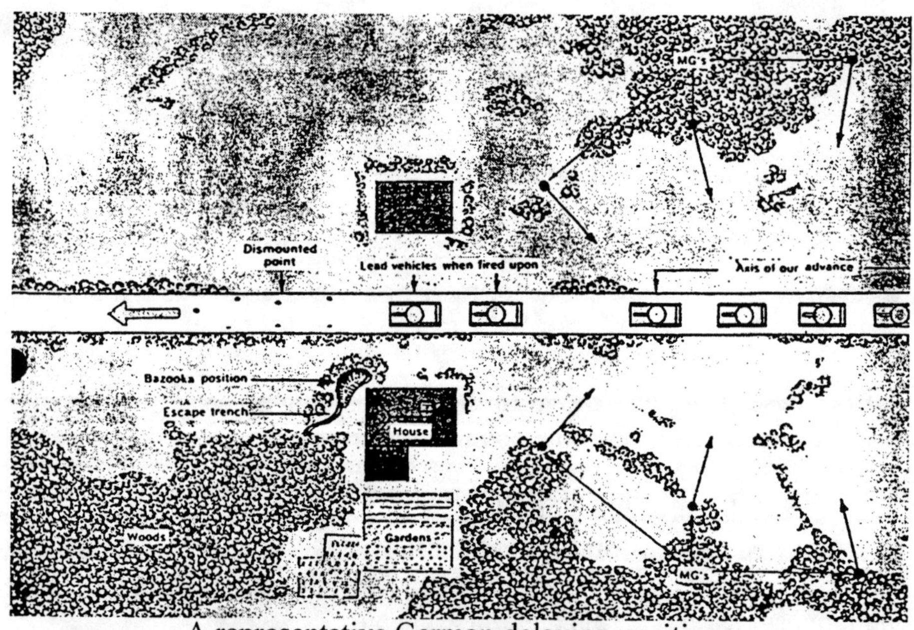

A representative German delaying position.

When the bazooka fires, all the machine guns open up on the remainder of the column, not so much to cause casualties as to create confusion and to make it difficult for the Allied force to tell the spot from which the bazooka fire has come. In fact, the Germans rely on a combination of factors -- the excellent camouflage of the positions and escape trench, besides the confusion created by the cross-firing machine guns -- to make it difficult for Allied soldiers to determine the points from which the resistance originates.

Intelligence Bulletin Vol II, No 12. (August 1944), pages 26-40

GERMANY:
TACTICS OF INDIVIDUAL GERMAN ARMS IN ITALY

The Battle of Italy has been primarily an infantry battle for the Germans. The machine gun, the mortar, and the mine have played parts of the greatest importance, chiefly because of the nature of the terrain, while the tank and the self-propelled gun have been obliged to undertake subordinate missions. This stress on the infantry arm, combined with a frequent need for the services of every available man, often has compelled the enemy to put men in the front line, regardless of branch. Engineers and reconnaissance units at times have been thrown into combat as ordinary infantry.

It has been a general German policy to commit only enough troops on the Italian front to block or delay the Allied advance. As a result, German commanders have had to use their strength very economically. After the Allied victory at Salerno, the Germans avoided committing a main force until Winter Line had been reached. Instead, they used highly mobile rear guards, flexible combat teams, and well situated defense areas -- all of which were characterized by economy of numerical strength and by generous allotments of automatic fire power. Counterattack on a large scale has been avoided, except to repulse penetration of a main line of resistance, and local counterattack usually has been undertaken only for the sake of delaying the Allied advance to some extent. This is the picture in brief. However, the work of individual arms warrants description in greater detail.

INFANTRY

In the early stages of the campaign, when the German withdrawal was conducted without much contact with Allied forces, German infantry was organized in small, mobile rear-guard groups. The composition and strength of these groups naturally varied considerably, according to the speed of the withdrawal, the extent of the delay that the Germans wished to impose, and the terrain. In general, however, the groups consisted of the motorized infantry or infantry in half-track vehicles -- equipped with a high proportion of light machine guns, in either case -- and often included support by tanks or self-propelled guns. Each rear guard included an engineer component, and sometimes a battery from the division artillery regiment. The basis of the rear guard is an infantry company. An infantry battalion fighting a rear-guard action normally sends only one of its rifle companies at a time on active missions. The three rifle companies are used in rotation, as long as their strength remains approximately equal. The following elements support the company (or companies, if the terrain makes it necessary to employ more than one): two or more antitank guns from the regimental antitank company, and half the heavy support weapons allotted to the entire rear guard -- that is, tanks, self-propelled guns, infantry guns, and gun howitzers. In country favorable to them, these reinforced infantry companies have proved capable of holding up a sizable Allied force on a fairly wide front.

When Allied pressure becomes strong, and disengagement consequently becomes more difficult, the single rifle company withdraws through the two remaining companies, which are supported by the remainder of the rear guard's heavy weapons. This leapfrogging procedure is continued until darkness approaches, when thinning-out takes place prior to a general disengagement. The withdrawal from one main position to the next, under cover of darkness, is an almost invariable procedure.

German rear guards in Italy withdraw by bounds to selected, but unprepared, positions. If it is the German intention to hold a line for some time, positions eventually are prepared.

During each stage of the withdrawal, individual company commanders can order retirement to the main rear-guard position, but only the commander of the main body can order withdrawal from one such position to the next. In the meantime the Germans make an effort to hold ground to launch counterattacks to regain vital features essential to an orderly retirement; inasmuch as withdrawal often must be conducted on a time basis, the enemy cannot afford premature retirement. The Germans launch large-scale counterattacks only when there is a threat to the main withdrawal or to the preparation of a main line of resistance, or when an established main line of resistance is in danger of being penetrated.

When a line is to be held for an extended period, German infantrymen take up a series of positions screening the main line and covering a network of observation posts. As far as possible, these positions are situated on forward slopes. Indirect fire is considered wasteful. Listening posts and outposts usually are established, to give warning of the approach of hostile forces. In the early stages of holding a line, wire and mines are not used. However, if further withdrawal seems unlikely -- for a time, at least -- mines and wire are used to give the forward positions additional protection. In such cases, the mines and wire are situated from 50 to 150 yards in front of the positions. Each of these positions, which are distributed fairly evenly over the company or platoon front, invariably holds two riflemen and two men and a light machine gun.

Heavy weapons, heavy machine guns, and mortars are sited behind the line of forward weapon positions. As a rule, the mortars are sited in pairs in the center -- on reverse slopes, if possible -- while the heavy machine guns are sited on the flanks. Where the field of fire permits, a mortar section may be strengthened by a pair of heavy machine guns. The heavy weapons remain under the battalion or company commander, depending on whether the battalion is Panzer Grenadier or Grenadier.

Dugouts for personnel and supplies are constructed to the rear of the forward positions, and are connected with the positions by communication trenches. (Whenever possible, the dugouts, too, are on reverse slopes.) It is interesting to note that the positions themselves generally are not connected with each other. Positions are lightly manned during the day -- with the machine gunners usually carrying the burden of defense, while the remainder of the personnel rest in dugouts. At night, forward positions are fully manned.

The screening positions are likely to be only a few hundred yards in front of what the German soldiers themselves regard as their main line of resistance (*Hauptkampflinie*). In static defense the distances between the forward positions, combat outposts, and so-called "main line of resistance" are greatly shortened. However, with the construction of switch lines (*Auffangstellungen*) to the rear, the main line of resistance tends to perform the work of combat outposts -- that is, to blunt the attack, while mobile elements, operating with the framework of the switch lines, counterattack and try to liquidate [the] penetration.

ARTILLERY

In the beginning of the Italian campaign, German artillery was principally engaged in covering infantry withdrawals and in delaying the Allied advance. For this function the Germans made extensive use of their self-propelled guns, which were employed so flexibly that they could be detached and assigned to rear-guard groups. The self-propelled guns had the mission of denying the use of roads,, bridges, defiles and so on to Allied forward units, so that infantry would be given a chance to retire to new positions. For this purpose the Germans made extensive use of 20mm antiaircraft-antitank machine guns. Moreover self-propelled guns were employed to cover road demolitions, and, in flat terrain, to form a mobile line of defense so that the infantry they supported could be concentrated on the main approaches. The self-propelled guns were committed in small numbers, often singly; they were well concealed behind walls or foliage, and frequently engaged targets at very close range. They were provided with infantry protection up to the time the infantry had to withdraw, and sometimes were employed to withdraw the infantry's heavy weapons, machine guns, and mortars, thus permitting the latter to fire until the last possible moment. When withdrawing as a battery, sections of self-propelled guns leapfrogged each other. As a result, one section always was ready for action while the others were on the move.

When the Italian front became stabilized, the self-propelled guns tended to fade out of the picture, except in support of raids and in local fighting, when it followed infantrymen and engaged strong points, machine gun nests, observation posts, and other objectives.

The field gun, on the other hand, played an increasingly important part in the campaign -- but not until after the early days of swift, evasive withdrawal, when tractor- or horse-drawn artillery had to move out well ahead of the infantry so as to have the use of the roads.

Basically, there has been no important change in German artillery tactics, although the current trends of the war, such as Allied air and matériel superiority, have brought about certain minor modifications. Targets are engaged in the customary ways. However, observation post officers often have to obtain the approval of battalion headquarters before firing, and barrages in the accepted sense of the term seldom are fired -- perhaps to economize on ammunition. The Germans are sensitive to Allied movement, and employ interdiction fire readily; but there seems to be no standard enemy thought as to which targets are most profitable. At critical moments the main target is the attacking infantry, and the Allied artillery receives only occasional fire.

Harassing fire is placed on areas affording defilade, and wherever the enemy has seen, or suspects, considerable grouping or movement. German harassing fire usually is carried out with a small number of shells of various calibers, and may be employed either by day or at night. Identification of Allied tanks or self-propelled artillery is likely to draw this type of fire.

Harassing fire is placed on areas affording defilade, and whenever the enemy has seen, or suspects, considerable grouping or movement. German harassing fire usually is carried out with a small number of shells of various calibers, and may be employed either by day or at night. Identification of Allied tanks or self-propelled artillery is likely to draw this type of fire.

Counterbattery work is left to medium and heavy units, because of their range and the destructive area of their projectiles. Long range firing sometimes is carried out without any attempt at precise adjustment. On the lower Garigliano and Anzio fronts, for example, shells are directed into fairly large areas known to contain guns and other targets.

The nature of the present campaign -- a planned withdrawal -- has enabled the Germans to register on all natural routes of advance and communication, as well as the most suitable sites for weapons, before their use by the Allies.

Allied air and artillery superiority, as well as effective counterbattery fire, has compelled the Germans to adopt several ruses to avoid disclosing their positions. They cease firing and halt all movement around their guns when hostile aircraft approach. To mislead attacking bombers, smoke shells are fired at short range when Allied smoke shells are indicating the positions to these bombers. Also, smoke is laid around positions to obstruct observation by hostile observation posts. Dummy flashes are set off to confuse flash spotting. Single guns or roving batteries are employed to fire from positions away from the normal battery sites, or to fire from the forward areas.

The excellent German camouflage shows that the enemy recognizes the need for concealment and deception.

Rocket projectors have been used, but only to a slight extent; the ammunition is almost invariably high-explosive. Smoke shells occasionally are fired from these projectors for screening purposes, but seldom are used for range estimation. Firing is chiefly indirect, involving the normal system for observation posts and forward observers. Positions are carefully camouflaged at all times, but are dug in only when the flash is hidden behind a crest and there is no need for an immediate move after firing.

ANTITANK WEAPONS

Antitank guns assigned to support rear-guard infantry companies are sited well forward and are employed with determination. As a rule, they are sited to the flank of good approaches and are concealed with great care. They tend to open fire at rather long ranges. Guns towed by half-track vehicles have taken part in infantry and tank attacks, in which they have supported the advance of the tanks. (The tanks have concerned themselves solely with the engagement of resistance holding up the infantrymen, and have left the neutralization of Allied armor to the antitank guns.)

The chief development has been the introduction of the antitank rocket launcher and hollow-charge antitank grenade. Both are infantry antitank weapons for use by company antitank sections in forward areas, either on approaches that tanks are expected to use or in the protection of headquarters. Ordinary weapon positions are dug on each side of an approach; the rocket-launcher uses one side, and the section leader with the grenade launcher uses the other. If the rocket fails to stop the tank, engaged at ranges of from 120 yards down to 60 yards, the grenade is brought into action at a range of about 35 yards.

Antitank sections of three rocket launchers sometimes are employed ahead of the forward infantry foxholes at such points as road junctions, and are sited so as to place fire on both approaches.

Because of the pronounced flash of the rocket launcher, which makes firing from a prepared position dangerous, and because of the splinter effect and flash of the antitank grenade, neither weapon appears to be too well liked by the individual German soldier.

As a result of the introduction of these weapons, there is a trend toward reorganizing the tank-hunting units in infantry companies.

TANKS

The subordination of tanks to infantry has been brought about by the nature of the terrain in Italy and by the general withdrawal plan adopted by the Germans. Also, the relatively small number of tanks available for combat in Italy has been a contributing factor. Tactics have been influenced by Allied air superiority. At Salerno, this air superiority forced the Germans to undertake tank attacks at night. Tank units were assigned sectors to which they were to confine themselves unless they were heavily hit; when this happened, they chose their own avenues of escape. (Air superiority also has forced the Germans to make all their movements at night, under cover of darkness.)

Operating exclusively in support of infantry, and with good coordination, tanks have been employed either in moderate strength, as at Anzio, or in twos and threes in rear-guard actions. The tanks move with the infantry, providing overhead covering fire for the troops in front, and protecting fire for the troops to the rear.

In open terrain German tanks often operate near buildings which offer the best -- sometimes the only -- concealment, as well as a certain amount of protection. In such circumstances they are likely to operate in pairs, to cover each other's movements.

Flame-throwing tanks have been used in close support of raids on strong points. These tanks have directed their primary weapon at personnel trying to withdraw from a position after it has received fire from other weapons. Regardless of whether the targets are personnel in woods, blockhouses, trenches, or ditches, the German intention is to drive them out into the open, where they will be more vulnerable to small-arms fire. Normally, the flame-throwing tank operates with other tanks but does not join in the action until the later stages. However, it may be used under conditions of poor visibility, when it tries to work its way close to a target without being detected.

ENGINEERS

As the *Intelligence Bulletin* has noted before, German minefields in Italy have been laid without much regard for definite patterns. Scattered mines are common. The intensity of antipersonnel mining is increasing, and so is the use of wooden-box mines.

Putting their knowledge of the terrain to good use, the Germans lay mines and set booby traps wherever the attackers are expected to advance or bivouac. Mines and booby traps have been found on beaches, at beach exits, in towns and villages, across roads and railways, in detours around demolitions, in road shoulders, in the spoil of craters, and under vehicle tracks; they have been found beside streams and along river banks, especially on the German side and near suitable crossings.

Antipersonnel mines have been discovered along hedges and walls, and various types of booby traps have been found in haystacks, ravines, and olive groves, on hillsides and terraces, and in valleys.

A wide variety of mines had been encountered, including Tellermines of all types, S-mines-Schumines, wooden box mines, concrete mines, and improvised mines.

The nature of the terrain has enabled the Germans to prepare many demolitions, which have been an essential part of the delaying actions and which have been executed with great thoroughness. Culverts and bridges have been destroyed completely. Roads and all suitable detours have been pockmarked with craters, blocked with abatis in the country, and blocked with the debris of buildings in towns and villages. Railway tracks have been blown up and ties cut. The debris left to obstruct movement is often mined. During periods when the front is relatively stable, German engineer units prepare demolitions to the rear. After a withdrawal, demolitions frequently are covered by snipers, machine guns, and self-propelled guns.

FLEXIBLE COMBAT TEAMS

Flexible combat teams, or "battle groups" (*Kampfgruppen*) have been prominent in the Italian campaign. Usually they are organized to perform some specific mission during the withdrawal; this mission may be to undertake a local counterattack or to defend a particular feature, the retention of which is necessary to an orderly execution of the movement. Such teams also have been used to plug gaps, to bolster sectors in which a threatening situation has developed, and to oppose Allied landings until a major force could be brought forward to counterattack.

The combat teams have varied in size from a company or two, with weapons attached for close support, to a regiment or several battalions, reinforced with tanks, artillery, engineers, and reconnaissance elements. Whereas the strength has varied, the types of elements have remained fairly constant. A combat team charged with conducting a rear-guard action is built around the infantry component, to which are added heavy infantry weapons from regimental companies, self-propelled artillery or a small number of tanks, and engineers. Antitank guns, antiaircraft guns, and -- less often -- field guns from division artillery also may be added.

Every effort is made to produce a balanced force. All combat teams include holding and support elements. Assault elements are added if an offensive action is contemplated.

DEFENSE AREAS

The Germans cover the lines of resistance or phase lines, marking the successive stages in withdrawal from one defense line to another, with a system of defense areas, or strong points. Just as it was the mission of the rearguards to prevent the pursuing Allies from making contact with the main German force and pinning it down, so the defense areas were established to prevent an Allied advance while a main German force was retiring from one position to another.

The German defense areas, like the rearguards, represent an effort to economize on strength. The typical composition in close country has been one or two self-propelled guns, a few heavy mortars, and as many as six machine guns. In more open country, small groups consisting of a self-propelled gun, two or three tanks, and a party of infantry (with machine guns) riding in personnel carriers have been encountered.

Defense areas usually are organized on the hedgehog principle. Although provision for all-around fire is made, defense areas are not necessarily mutually supporting. They generally are established on commanding features -- and sometimes on the forward edges of villages, if these command defiles. However, the Germans seem to feel that villages on flat terrain are too susceptible to artillery fire; for this reason, the enemy is more likely to establish defense areas to the rear of such villages, to engage the advancing forces as they debouch. Positions are changed frequently. In hilly country the Germans have used these defense areas extensively to force considerable deployment and subsequent full-scale attacks; however, it is a favorite enemy tactic to slip away just before the attack materializes.

Intelligence Bulletin Vol III, No 4. (December 1944), pages 46-52

GI COMMENTS ON GERMAN USE OF FIRE POWER

As the war in Europe progresses, U.S. soldiers are becoming increasingly familiar with the ways in which the enemy employs his fire power; however, since each unit naturally has learned more lessons from its own experiences than from those of other outfits, a general pooling of information can be extremely helpful.

Most of the following comments on German use of fire power have been submitted by U.S. noncoms and company officers, and are based on fairly recent combat experiences in France and Italy.

Artillery in Support of Infantry

"The German basic force seemed to be infantry with heavy weapons, plus a heavy tank, a self-propelled gun or some other single cannon. The enemy covered an attack by an impressive display of fire power over a wide front. He also attempted to move his base of fire by having a self-propelled gun or a tank accompany the infantry, who fired machine pistols as they advanced. In defense a single gun was used, but we destroyed so many that the Germans soon found this system wasteful. To gain control, they stayed in close columns very near the front where we frequently surprised them."

"The chief drawbacks of German field artillery can be summarized briefly as lack of mass, poor transport, poor lateral communication (especially between observation posts), slowness in occupying positions, and preparing for massed fire, and reliance on single cannon, often very poorly sited, for support."

"When German delaying forces used cannon, they tended to select poor positions. Their high-velocity weapons had to occupy positions which we could discover easily. Their infantry howitzers were very poorly situated, either through ignorance or because of a desire to put the guns in a position from which their personnel could not escape and where they would have to fight to the last."

"On the other hand, we were impressed with the accuracy of German field artillery. I've seen a 1`50mm battery concentration hit a crossroads so consistently that engineers had to be called on to make it passable for a 2 1/2-ton truck. As far as thoroughness goes, the Germans get more out a round than the devil himself gets on a lump of coal."

Deceptive Fire of Machine Pistol

"The German machine pistol -- a submachine gun, in U.S. terminology -- is very deceptive when heard in combat. We call it the 'zipper gun.' It can be fired almost on top of you and yet sound far away, and visa versa. The reason for this is that there is a gadget on the barrel, which enables the operator to muffle the sound of firing. Our men know that this gun is not highly accurate, but, because of the high rate of fire and the sound, it's a fairly harassing weapon."

"The first bursts that the machine pistols delivers are effective. The rest have a tendency to go high and to the right. Because of this, new troops sometimes think that they are being fired on by more weapons than are actually in operation. A favorite German trick is to fire a single round from the machine pistol, move to another spot and deliver automatic fire, and then move again to fire a single round."

Prearranged Fire on Approaches

"When the Germans expected us to attempt an advance, they would zero in on all the avenues of approach that we might reasonably be expected to use. Then they would plant snipers at strategic points. When we attacked, the snipers would open up immediately, in an effort to pin us down into a compact group. If this tactic succeeded, the enemy would let loose with mortars or 88's, which already had been zeroed in. The Germans especially favored this method in hilly terrain, where we did not have a wide choice of attack routes."

"The Germans seldom defended low ground. They almost invariably entrenched themselves in high ground, where they would have good observation."

Artillery Fire Control

"In our experience, German artillery fired only on targets of importance, except in the case of a diversionary attack, when firing at random seemed to be a general practice. The Germans fired according to the amount of ammunition they had on hand. To conserve ammunition for their larger guns, they would couple one of these guns with a light gun, and would try to obtain the range using the lighter piece (However, by using smoke, we were able to thwart these efforts at range estimation.) The Germans invariably used only one destructive weapon against a target. When they were hard-pressed and were about to withdraw from a position, their artillery would fire a heavy barrage, lasting as long as half an hour, on the whole front or area. This barrage would be heavier than in the case of an actual attack. "

"Our unit found that the Germans nearly always fired a smoke shell during the daytime, to get the range for their artillery. At night they would move a machine-gun squad close up to the line, and fire tracer over a certain point; by this method the observation post could determine the range for artillery at night."

"At first the Germans used only two men to a machine-gun nest, but later on they began to use three men. The third man would stay hidden in case there should be an Allied attempt to take the machine-gun nest. If such an attempt was made, the two German soldiers who were visible would walk out in front of the machine-gun nest, holding up their hands to be searched. When Allied soldiers were engaged in the searching process, the two Germans would suddenly drop to the ground, and the third German, concealed in the machine-gun position, would start firing."

"The two Germans suddenly would drop to the ground, and the third German, concealed in the machine-gun position, would start firing."

"When the enemy was driven out of a town or village, he would leave an observation post in a church tower or some other place where there was a large bell. As our troops entered the village, the observation-post personnel would ring the bell, and German artillery, having zeroed in previously, would fire as soon as the bell rang."

Use of Tanks

"I never saw a German tank employed singly. In nearly all instances, a section or a platoon was employed. One tank may try to draw your fire; then, if you react as the Germans expect you to, you are immediately subjected to the remainder of their fire power."

"German tanks have a tendency to bunch up, and it is quite common for them to expose their broadsides. We found them vulnerable to cross fire from firepower employed on an extended front."

"I found that the enemy employs his tanks in groups of six or more, and that there usually were two or three types in a group. The most common, we found, were the Pz.Kw. VI, the Pz.Kw. IV, and, in most cases, one or two self-propelled guns. These guns, I believe, are intended to delay our advance in the event that the tanks have to withdraw or maneuver to a more advantageous position. The Germans frequently use a single tank as a decoy to draw your fire, with the hope that you will present yourself as a more vulnerable target. The enemy's main fault, it seems to me, is bunching up his vehicles, and trying to get too much through a single avenue of approach or withdrawal."

"If a German tank is not completely destroyed -- set afire with high-explosive shells, for example -- the enemy is likely to sneak back into it and deliver unexpected fire from its weapons. Also, a crew bailing out may leave a man behind to cause us trouble. Once, we fired on a Pz.Kw. IV Special, and hit it in the track. The crew bailed out immediately, and we thought the tank was out of action. However, the gunner remained in the vehicle. After we had stopped watching this particular tank, the gunner fired two rounds at us. We weren't hit, fortunately, and lost no time at all in demolishing the tank."

The crew bailed out immediately, and we thought the tank was out of action. However, the gunner remained in the vehicle. After we had stopped watching this particular tank, the gunner fired two rounds at us.

Antitank Guns

"The Germans have been introducing more and more 75mm antitank guns, usually emplaced in pairs. The positions are likely to be just below the crest of a hill or high bank. From the positions of the guns, it is evident that the crews plan to let our armor come well within range, and then take them under a cross fire. The emplacements are always well dug-in and camouflaged. Invariably there is a crawl trench leading from the gun itself to a dugout, which serves as living quarters for the crew. After the gun has been hit, surviving members of the crew move down these trenches and out of observation."

Machine-gun Fire

"Our men have learned how to get around the fast-shooting German light machine guns. These guns have such a rapid rate of fire that they are not able to cover a great deal of ground. When our men have stayed well apart, the machine guns have not been able to do much damage. Actually, these weapons are terrific ammunition wasters. And our men have learned how to take advantage of the few moments afforded when the crew must change barrels. This happens frequently because of the high rate of fire. What ground the light machine guns cover is covered well, but it is a very limited area."

"Double rows of German base fire, at night, involved a heavy unidentified line of fire approximately 3 feet from the ground and a high, arching line of fire, amply identified by tracer bullets. Evidently the Germans hoped to create the impression that the principal fire was high and inaccurate, and also to discourage night bayonet attacks. "

Intelligence Bulletin Vol III, No 7. (March 1945), pages 48-53

IN BRIEF

BICYCLE MOUNTED TROOPS

Newly created Volksgrenadier divisions not only have a bicycle-mounted reconnaissance battalion or company, but also have an entire battalion of infantry mounted on bicycles. In addition, the two engineer companies of the division Engineer Battalion are bicycle-mounted. It may be assumed that some of the tactics employed by this bicycle-mounted company in the reconnaissance unit (*Fusilier Batallion*) of the infantry division may also be used by the bicycle-mounted of the Volksgrenadier divisions. Here are several prisoner-of-war comments on this subject.

A prisoner remarks that when a bicycle-mounted squad is moving along a road as a point, anticipating contact with a hostile force, the squad leader and a runner are followed at a distance of about 50 yards by three machine gunners with a light machine gun, supported by a sniper, a semiautomatic rifleman, and two riflemen, one of whom is armed with a cup grenade discharger. When the squad is fired on, the machine-gun detachment immediately deploys, while the remaining men drop their bicycles under the nearest available cover and take up firing positions.

The leading squad of a platoon is said to move with a rifleman, a semiautomatic rifleman, a machine gunner with a light machine gun, a sniper, the squad leader and a runner, two machine gunners, and a rifleman armed with a cup discharger -- moving in that order. Fifty yards behind, the platoon commander and a runner, the platoon sergeant and a runner, a telegraph operator and a medical aid man, and an antitank rifleman follow -- in the order named.

A prisoner from another unit comments that in his outfit it was common practice to send two bicycle-mounted scouts ahead of the point squad.

Prisoners remark that bicycle-mounted companies are expected to be able to cover up to 75 miles a day, but that, in actual operations, the figure seldom exceeds 50 or 60 miles.

Prisoners from certain bicycle-mounted companies say that they have been trained mainly in infantry tactics, and not primarily for reconnaissance missions. One unit was trained to move forward on its bicycles, leave them in farm buildings, and then go forward on foot to fight as infantry.

In Russia a company was detached from an infantry regiment, equipped with bicycles, and formed into a reconnaissance company. These men were given the mission of protecting the regimental flank upon contact with a hostile force.

ASSAULT GUN TACTICS

To teach German infantry some of the tactics used by assault guns, the Fifteenth German Army outlined the advantages and disadvantages of these self-propelled weapons so that the infantry could have a better understanding of how to cooperate with them in the field.

In reply to the question, "What must the infantry know about the assault guns?" the Germans offer these comments:

"The assault guns are the strongest weapons against hostile tanks. They will engage all your most dangerous enemies, and destroy them or force them to take cover. Assault guns are strong when concentrated, but have no effect when used in small numbers. They are capable of forward fire only, since they have no turrets; therefore they are most sensitive to attack from the flanks. This is why the guns must never be employed by themselves, but always in conjunction with infantry. These weapons may be considerably restricted in marshy land, thick woods, and natural or artificial obstacles; moreover, they constitute large targets. They can see and hear little. Even during a battle, the assault guns occasionally must withdraw to cover, and obtain fresh supplies of ammunition and fuel."

"This brings us to the question of how the infantry should assist the assault guns."

"Infantry must draw the guns' attention to hostile tanks and other targets by means of the signal pistol, prearranged light and flag signals, and shouting."

"Infantry must neutralize hostile antitank guns."

"The flanks of assault guns must be covered and protected by the infantry against hostile tank-hunting detachments, which are always ready to operate against our assault guns. Such protection is especially necessary in built-up areas and in terrain where visibility is poor."

"The infantry must warn the assault guns of the proximity of anti-tank obstacles and mines, and must be prepared to guide the guns through such obstacles."

"The infantry must take advantage of the guns' fire-power to advance in strength via prearranged lanes not under fire."

"The assault guns must be given sufficient time for reconnaissance. The guns and the infantry will formulate plans through personnel consultation, and will ensure means of communication during battle."

"Infantry should not stay too close to the guns, and should not bunch. Instead, deployment is advised, to lessen the danger of drawing hostile fire and to avoid injury by ricochets."

"Since the driver of an assault gun has limited vision, infantrymen must keep in mind the danger of being run down, and must move accordingly."

"Assault guns are 'sitting' targets when they have to wait for the infantry; infantry can find cover almost anywhere, but the assault guns cannot."

"Since the guns fire at the halt, the infantry must gain ground while the guns are firing."

"Although the assault guns are of great assistance when ground is being gained, it is the infantry that must *hold* the ground."

"Since the assault guns must keep their ammunition available for unexpected or especially dangerous targets, the infantry must engage all targets that it can possibly take on with its heavy and light weapons."

"Although the assault guns must withdraw after every engagement, to prepare for the next engagement where their assistance will be required, the infantry will *not* withdraw."

THE LIGHT MACHINE GUN IN FOREST FIGHTING

The comments of a German company commander on the use of the light machine gun in forest fighting on the Eastern Front are worth noting. In German training, this officer says, it is always emphasized that in forest fighting the rifles are forward, with the light machine gun in the middle or rear of the squad. However, after commanding a company during 4 weeks of forest fighting, the officer decided that it was preferable to have the machine gun forward. He remarks that practically all forest fighting is the equivalent to an assault operation, and that a German infantry training manual observes, in this connection: "Just before and during the break-in (*Einbruch*), the hostile force should be engaged at the maximum rate of fire by all weapons. The light machine gun will take part in the assault, firing on the move, alongside the rest of the squad." In forest fighting, however, the danger from ricochets is so great that the machine gun definitely should be forward, this enemy officer says, and adds that when his company had learned this lesson, it had considerable success in avoiding casualties by keeping the weapon well forward. He cites an example of the revised tactic:

"My company was making an attack in a wood where a Soviet force had dug itself in, in a number of positions prepared in depth. The first firing failed to make the opposition reveal itself. My company was baffled at being held up in this manner. The machine guns then were brought forward, and, firing continuously from the hip, my company stormed in with spirit to take the first position. The fire forced the Russians to hit the ground, and gave them no opportunity to put down aimed defensive fire. Our casualties were practically all from mortar fire."

"Having the machine gun forward has one further advantage. In woods, squads often get split up and scattered. If the machine gun is forward, however, the squad can easily rally around the machine gun, and, after taking a hostile forward position, is able to reorganize speedily and go on to take the enemy's second and third positions."

"We found that it was not necessary, as the Infantry Training Manual suggests, to make the assault at the double. As a matter of fact, my men were in no physical condition to do so. In this type of fighting, we always advanced at a walk."

"I also found that my men advanced more confidently, since, because of the noise of the machine guns and their own cheering, they could not hear the Soviet bullets. In defense, too, it is highly advisable to encourage the men to cheer as the enemy attacks."

From a high German echelon come the following comments on the company commander's observations.

"1. When advancing in woods and in close touch with hostile force, machine carbines and rifles (preferably the automatic rifles) will be detailed to scout and provide close protection forward of the squad. The machine gun therefore will move with the squad, and not ahead of it. When the machine gun goes first, it is detected too readily and the detachment is fired on before it can bring the weapon into action."

"The more dense a wood is, the more the conditions of fighting approach those of close combat; that is to say, machine carbines, automatic rifles, and egg hand grenades play a decisive part."

"2. As soon as a hostile force is encountered, the machine gun will be employed forward, for the reasons that this company commander has given in his report."

"The usefulness of the machine gun here lies chiefly in the fact that, like the 'cheering,' it helps to keep morale high."

"When a hostile force is well dug-in, and is thoroughly prepared to meet an attack, the initial crack in the opposition's front must be made by assault operation."

DISCOVERED IN COMBAT

X MARKS THE SPOT

"In France the Germans frequently painted 6-inch yellow crosses on tree trucks. When viewed through binoculars, those crosses were easily visible at a distance of 700 yards. We discovered that the yellow markings were used as zero points for the German 88s, and thereafter took care to give any such potential target a wide berth."

VULNERABLE STRONG POINTS

"In December my company was near Harlange, in Luxembourg. One night I sent out reconnaissance patrols of four and five men. They brought back information that the Germans had established four strong points in front of my company's position. These strong points were about a quarter of a mile apart, and each contained 40 to 60 Germans. There were no German soldiers between the points. Each point was equipped with six or seven machine guns, and with machine pistols, rifles, and bazookas; in addition, a couple of the points had 80mm mortars. Patrol members reported that the Germans had posted only two sentries, one at each end of the strong point series. Upon receiving this information, I assigned one platoon to attack each strong point. During the night these platoons crossed 550 yards of open ground to reach the strong points. They crept within close hand grenade range, and waited for dawn. As soon as there was light enough to see by, each platoon attacked the German position with hand grenades. One point was completely destroyed. In another, only three of the enemy survived. The Germans in the other two points fled, and I regret to say that a number of them succeeded in escaping. The important thing, it seems to me, is that this entire action was because the Germans had been lax in posting adequate security."

"Earlier, the Germans themselves had used similar tactics against another company to the right of my own, and had inflicted severe casualties on our men. In this action the Germans used flashlights to locate our men in foxholes."

BOTTLE MINES

"The Germans in France improvised 'bottle mines.' Usually these were made of large wine bottles. The German soldiers filled the lower half with earth, and the upper half with a yellow crystalline explosive mixed with bits of copper wire, nails, fragments of tin, and so on. They then corked the bottle with a Z.Z.42 fuze, and laid the contraption as an antipersonnel mine with a trip wire."

SNIPING AT LEADING ELEMENTS

"We found that the Germans usually tried to pin down our leading elements by directing fire from machine pistols and other small weapons against their front and machine-gun fire against their rear. The Schmeisser machine pistol has a high cyclic rate of fire, but is by no means accurate. It was used extensively by German snipers who placed themselves 3 or 4 miles outside of towns or villages, along the roads leading to these communities. The snipers would cut in on the leading element of a company or battalion in order to hold it back. They would fire until they were out of ammunition, and then would jump out of the trees and coming running toward our lines, shouting *'Kamerad!'* Each sniper wore a specially camouflaged uniform, and also had camouflaged his weapon by painting it and tying leaves to it so that it would blend with the surrounding foliage. An important mission of German snipers was to delay the advance of our columns into populated places. This was why they fired on our leading elements, instead of holding their fire and trying to engage a larger force."

The Schmeisser machine pistol .... was used extensively by German snipers who placed themselves 3 or 4 miles outside of towns or villages, along the roads leading to these communities.

MASKING THE 88

"About 40miles north of Florence, my outfit captured towed German 88mm guns. We had observed one of these guns firing, before we captured it. The guns had been spaced about 1 mile apart, and had been hidden with the enemy's customary care. The rear walls of two houses had been removed, and the inner side of the front walls had been reinforced with cement, making the total thickness of each front wall about 5 feet. An 88mm gun had been emplaced in each house, with the muzzle behind an open window in the front wall. As a result, the guns were well hidden, and the guns served as smoke shields."

SUICIDE PATROLS

"Recently the enemy has been sending out suicide patrols, each consisting of about six men armed with machine guns. The mission of these suicide patrols has been to infiltrate our lines and do as much damage as possible. We generally succeed in wiping out such patrols on contact, but they can be something of a nuisance, even so. One good thing about it is that when you liquidate a suicide patrol, you feel that you're getting rid of Nazis of the most fanatical type."

"One of the favorite tactical measures that Germans employed in Normandy involved the shifting around of their antitank guns at night. All their antitank guns were mobile, and at night the Germans sometimes moved them to positions around the perimeter of an area occupied by U.S. tanks. At the break of dawn, the Germans would lay down a smoke screen to cover their positions, and then would open fire on the U.S. tanks within the perimeter. A favorite position for the German antitank gun was a 'crossrow,' where hedgerows intersected at right angles. Before a U.S. force advanced, the German antitank gunners would draw a bead on a particular object in the distance -- a tree or a signpost, for example. Later, when a tank or any other vehicle came near that object, the German guns could open up and have the range correct with the first round. German marksmanship with artillery pieces and antitank guns was excellent, although their marksmanship with small arms was inferior to ours."

———

DEFENSE AGAINST PATROLS

"German machine-gun nests frequently would detect our 10-men night patrols approaching, but would refrain from opening fire. Instead, the Germans would let the U.S. soldiers sift past them. When the patrol had advanced about 50 yards beyond the machine-gun nests, the Germans would open up with everything they had. Our men, of course, were not only confused by this unexpected fire from their rear, but suffered casualties. They learned to conduct even more extensive observation by daylight, in an effort to detect every possible point which might harbor a machine-gun nest."

When the patrol had advanced about 50 yards beyond the machine-gun nests, the Germans would open up with everything they had.

———

ARTILLERY SIGNAL

Although the Germans have not been reported as making wide use of the following trick, it seems worth mentioning, if only because variations of it may possibly be encountered in the future:

"While the First Army was chasing the Germans across France, our outfit arrived at a small town after dark. This town had been evacuated by the Germans just a few hours before. They had been forced to abandon a number of motor vehicles, and the headlights on one of those vehicles had been left burning. When one of our men turned off the headlights, so as not to disclose any of our activity, the Germans began directing artillery barrages on us. It was clear that the artillery had been zeroed in on the area where the vehicles were, and that by extinguishing the headlights we had fallen in with an enemy plan and had indicated our arrival in the town."

Intelligence Bulletin Vol III, No 6. (Feb 1945), pages 19-23

DISCOVERED IN COMBAT (non-tactical articles were excised)

REACTION TO ARTILLERY FIRE AGAINST TOWNS

"When U.S. heavy artillery destroyed buildings -- even fortified buildings -- in Brest, without making sure that the direct support artillery could maintain neutralization until the infantry assaulted the area, the Germans made the most of their opportunity when our fire was lifted. They promptly moved back in, and constructed fortifications from the rubble. When the Germans did this, their new positions often were harder to reduce than the original buildings would have been."

———

"German troops in Brest who were provided with adequate cover seem to have been affected only slightly by intermittent harassing fire, even when the fire was from heavy artillery. As soon as the men became convinced that their cover afforded reasonable protection, occasional rounds failed to disturb their normal routine."

———

COMBAT IN TOWNS

"At no time while our outfit was engaged in mopping-up operations in Aachen did the enemy fire a shot from behind our lines. As we went along, we searched every room and closet in every building, and blew every sewer which might have afforded the enemy a hiding place. Not only were our fighting men relieved of the fear of being sniped at form the rear, but command and supply personnel functioned more efficiently."

INFANTRY-TANK NIGHT ATTACK

"On one occasion the Germans launched a tank-infantry attack at night against our positions -- and over muddy ground. After taking the objective, the tanks withdrew before daylight, leaving their infantry to hold the ground. A counterattack restored our position."

"In the attack the Germans sprayed the area with fire, and used star shells and flares, in an attempt to frighten our troops. The enemy tanks didn't stick to the roads, but maneuvered across country, racing their engines and milling around to cause confusion among our infantry. Our infantry fired machine guns in the direction from which the sounds of the tanks came, and the sparks of the ricochets located the vehicles sufficiently to permit the tank destroyers to fire. Incidentally, a German self-propelled gun was knocked out as a result of this activity."

VULNERABILITY OF PILLBOXES TO "SEALING"

"The amount of TNT needed to blow German pillboxes can be reduced considerably if the escape hatches can be found and plugged beforehand. These hatches, which are encountered in nearly all pillboxes, are about 2 feet square. They are likely to be plastered over and hard to detect."

"Pillboxes have been vulnerable to effective demolition when charges have been placed in the ventilation pipes, which run vertically through the side walls near the pillbox entrance. First the bottoms of the pipes are plugged, then 30 to 50 pounds of TNT are dropped in, primed, and tamped. In one instance, firing the charge breached the wall completely, and the surviving occupants were either wounded or stunned by flying concrete."

"Embrasure openings have been obstructed by means of thermite grenades. If the sliding door of the embrasure is closed, a grenade is placed on the slideway, is activated, and becomes a molten mass. Although the door itself is not welded, it is jammed by the mass, which hardens and thus obstructs the slideway. A single grenade is sufficient to jam a small door, but two grenades are used against large doors with armor plate more than 2 inches thick. If the grenade cannot be placed on the slideway, a trough of ¼-inch metal may be used, to cause the molten mass to run into the slideway. The surface on which the weld is to be made should be dry and clean. If the door of a German pillbox works on hinges, jamming cannot be accomplished by means of thermite grenades, since the molten mass cannot be controlled sufficiently to create a strong band between the door and the frame.

Intelligence Bulletin Vol II, No 4. (December 1943), pages 31-44

Section I. GERMAN MACHINE GUNS AND NOTES ON THEIR USE

1. INTRODUCTION

This month the *Intelligence Bulletin* devotes two sections to timely information regarding German machine guns – a subject to particular interest to junior officers and enlisted men. Readers who wish to investigate this topic more extensively are referred to two M.I.D. publications: "German Infantry Weapons" (*Special Series*, No. 14) and "The German Squad in Combat" (*Special Series*, No. 9). The former is of value for its abundant technical detail, while the latter contains useful information about the tactical employment of a machine gun by the squad.

Two types of German-manufactured machine guns are used by the German Army. These are the 7.92mm MG34, which was introduced before the present war, and the 7.92mm MG42, which was introduced in 1942 and is gradually replacing the MG34. Both types are issued for use in the following roles:

    a. As a light machine gun, fired from a bipod.

    b. As a heavy machine gun, with tripod and a telescopic sight.

    c. As an antiaircraft machine gun, fired from single and twin antiaircraft mounts.

Also, the MG34 is mounted on nearly all German tanks and armored cars. German machine guns normally remain in use in the roles for which they are first issued. Production, maintenance, and training are simplified, however, by the Army's adoption of a single standard model. (Any existing duplication is caused by the present change-over period.) Ammunition is interchangeable throughout.

The issue of ammunition to infantry is roughly as follows:

Light Machine Guns:

Ball	84%
Armor-piercing	12%
Armor-piercing tracer	4%

Heavy Machine Gun

Ball	82%
Armor-piercing	10%
Armor-piercing tracer	8%

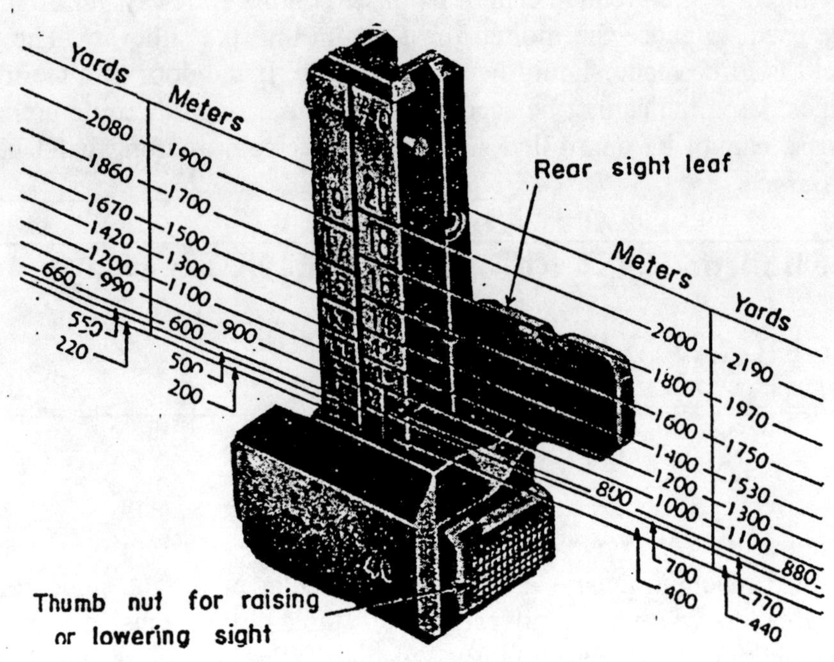

Figure 6. – Rear Sight of MG35 (showing relation between yards and meters.

2. THE MG34
a. Table of Characteristics

Weight (unmounted)	24 lbs
Weight of bipod	2½ lbs
Weight of HvMG tripod	42 lbs
Weight of LMG light tripod	14 lbs 11 oz
Over-all length of gun	48ins
Cyclical rate of fire	800-900 rpm
Practical rate of fire (LMG)	150 rpm
Practical rate of fire (HvMG)	300 rpm
Cooling	Air
Cartridge feed	Flexible metal belt containing 50 rounds (two or more of these may be joined end to end) or a drum containing 50 rounds.
Ammunition carriage	Belts carried in metal boxes; weight, with 300 rounds, 22 lbs.
Sights	Blade front sight and leaf rear sight graduated from 200 to 2,000 meters (see fig. 6). Telescopic sight when used as a HvMG.

b. Method of Operation

The gun is recoil-operated (by a barrel recoil of ¾-inch.) This action is assisted by muzzle blast. The breech mechanism is of the Solothurn type (rotating bolt head). Provision is made for semi-automatic fire.

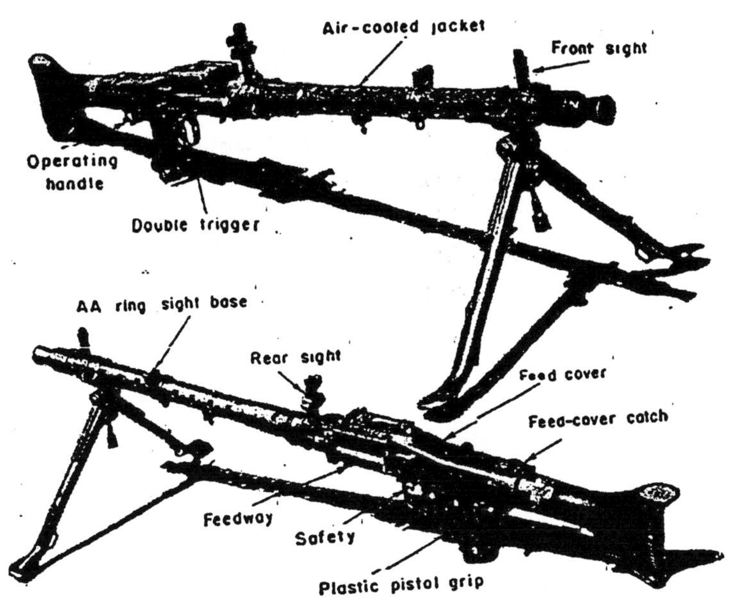

Figure 7. – Two views of the MG35 on Bipod Mount.

c. Use as Light Machine Gun (See fig. 7.)

As a rule, the gun is fired from the bipod (see fig. 8), although three very light tubular tripods are   carried by the company and are issued to the platoons as required. These light tripods are used chiefly in antiaircraft defense, but occasionally terrain conditions may warrant the use of the gun from this tripod against ground targets.

The open sights are graduated to 2,000 meters, but German manuals give the maximum effective range as 1,500 meters (1,640 yards). The most effective range seems to be between 650 and 850 yards.

The barrel must be changed after 250 rounds of more or less continuous fire. Two separate barrels are carried in separate barrel cases by members of the light machine-gun detachment.

Bursts of 7 to 10 rounds are fired, and approximately 15 aimed bursts can be fired in 1 minute.

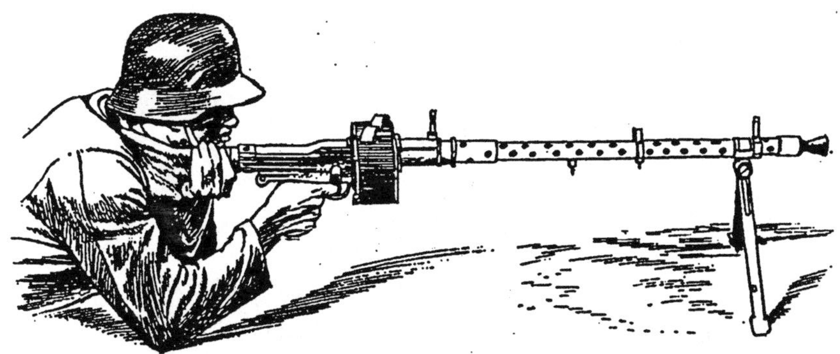

Figure 8. – German Method of Firing MG35 fr4om Bipod Mount.

d. Use as a Heavy Machine Gun

When the MG 34 is used as a heavy machine gun, it is mounted on a tripod known as the MG *Lafette* 34 (see fig. 8). The gun is carried in a cradle embodying a spring bugger, which enables the gun to recoil as a whole on its mount. This devide adds greatly to the stability of the mount without increasing the weight. Adjustable elevating and traversing stops are provided; these enable the gun to be elevated and traversed within predetermined limits.

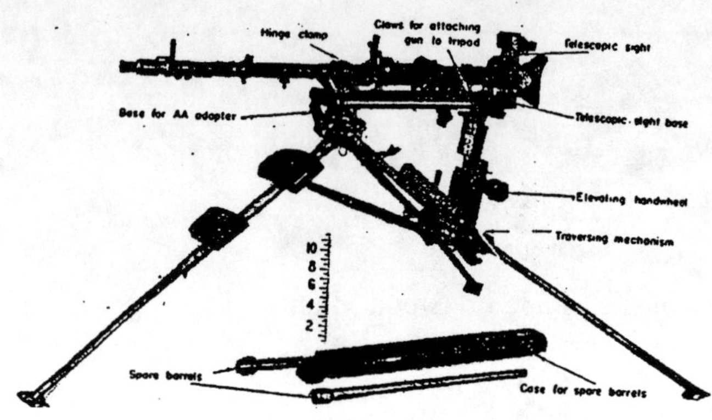

Figure 9. – MG34 on Tripod Mount.

An automatic searching fire device is incorporated and is operated by the recoil of the gun in the cradle.

A telescopic sight is fitted, with provision for direct fire up to 3,250 yards and for indirect fire to 3,800 yards. Small targets are not engaged at ranges of more than 1,500 yards. Three spare barrels are carried.

e. Use as Antiaircraft Machine Gun

Four types of antiaircraft mounts are provided. These are (1) the standard light tubular tripod issued for the light machine gun (see fig. 10); (2) the standard heavy machine-gun tripod with an adaptor; (3) a monopod mount fixed in vehicles used for transporting personnel; (4) A twin mount carried in a small horse-drawn, two-wheeled trailer. The standard tripods mentioned in (1) and (2) can rapidly be converted for antiaircraft use.

Figure 10. - MG34 on Antiaircraft Mount (using drum feed).

German soldiers are also taught to fire on aircraft and ground targets by supporting the gun on another man's shoulder. Regardless of the method used, fire above a maximum vertical range of 2,600 feet is prohibited.

f. Use in Armored Vehicles

Nearly all German tanks and armored cars are armed with one or two MG34's, in addition to their main armament. Under these circumstances the butt is removed, and the gun is fixed in a special mounting in the gun.

3. THE MG42
a. Table of Characteristics

Weight (unmounted)	20 lbs
Weight with bipod	20¾ lbs
Weight of HvMG tripod	43¼ lbs
Weight of LMG	14 lbs 12½ oz
Over-all length of gun	48 ins
Cyclical rate of fire	1,100-1,150 rpm[1]
Practical rate of fire (LMG)	150 rpm
Practical rate of fire (HvMG)	300-400 rpm
Cooling	Air
Cartridge feed	Flexible metal belt containing 50 rounds (two or more of these may be joined end to end).
Ammunition carriage	Belts carried in metal boxes; weight, with 300 rounds, 22 lbs.
Sights	Blade front sight and leaf rear sight graduated from 200 to 2,000 meters. Also uses telescopic sight on tripod when employed as a heavy machine gun.

b. Method of Operation

The principal of operation is a combination of recoil and blow back. In place of the Solothurn rotating bolt-head action of the MG35, there is a new system. This involves a lateral separation of the bolt studs in the cylinder from the bolts in the barrel extension. An improved feed mechanism is provided. Barrel-changing is extremely rapid. No provision is made for firing single rounds.

c. Construction

The extensive use of stamping, riveting, and spot welding gives the gun a less finished appearance than that of the MG34. There are few machined parts. However, this does not mean that its life is shorter or its performance is inferior.

---

[1] German documents give the cyclic rate of fire as 1,500 rpm.

Use of MG35 and the MG42 as light machine guns is comparable, except that, in the case of the newer model, the higher rate of fire and the consequent "creep" of the gun makes shorter bursts (of 5 to 7 rounds) advisable. Twenty-two aimed bursts can be fired in 1 minute.

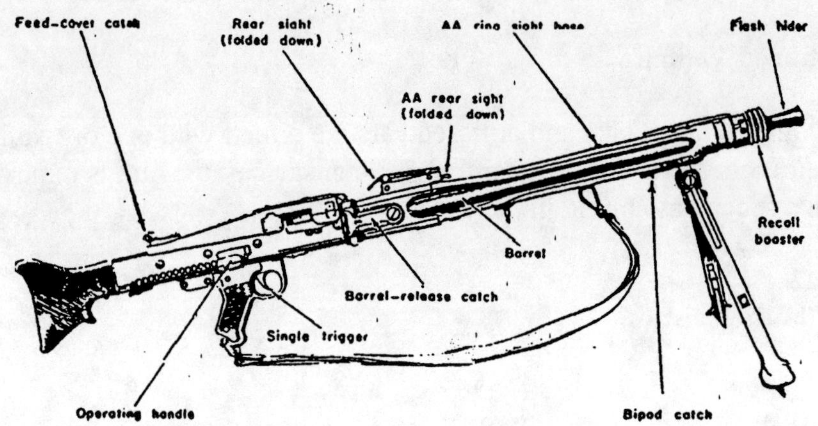

Figure 11. – MG42 on Bipod Mount.

e. Use as Heavy Machine Gun

The tripod has been slightly modified to suit the catches on the gun, but has not been altered in essentials. A new and slightly modified telescopic sight is also in use.
German manuals advise bursts of 50 rounds for best results.
As in the case of the MG34, the barrel should be changed after 250 rounds of more or less continuous fire, although prisoners of war have stated that barrels need not be changed until 400 rounds have been fired. It has been suggested that the detachment may carry more than 3 spare barrels, but this point has not yet been established.

f. Use as Antiaircraft Machine Gun

Against aircraft the MG42, like the MG34, is fired from light and medium tripods as well as from single and twin mobile mounts.

g. Possible use in Armored Vehicles

Although use of the MG42 in tanks or armored cars has not yet been reported, it is being used in the "Ferdinand," and further development along this line should be regarded as a distinct possibility.

4. GERMAN TACTICAL DOCTRINE
a. General

(1) *Light machine gun.* – While the basic principles of the tactical use of the German light machine gun are not unusual, the following are especially stressed in enemy doctrine: surprise, fire and movement, coordination of fire power, conservation of ammunition, and alternate positions.

(2) Heavy machine gun. – When issued with the heavy tripod, the MG35 and

the MG42 are invariably used as heavy machine guns, except in the case of surprise attack by a hostile force, when a gun may be off its tripod and not in a firing position. Because of its dual nature, the weapon can be used as a light machine gun under such circumstances.

In the German Army the heavy machine guns are carried in the machine-gun company of the infantry battalion. This company has twelve heavy machine guns in addition to six 3-inch mortars.

The heavy machine gun is employed from open or covered positions. In the case of open positions, which may be taken up in battle, use is made of all available cover. A position of this kind is normally manned by a section (two guns) under the control of the section leader. Covered positions are generally on reverse slopes, and are normally manned by a whole platoon (four guns), which the platoon commander controls from a central command post.

The heavy machine gun is prepared for action behind cover, and is placed in its firing position only at the last moment.

Covered positions are almost always used when overhead fire is to be delivered.

b. Attack

In the attack heavy machine guns cover the deployment of the rifle companies from echeloned positions sited on commanding ground. In a penetration (*Einbruch*) the heavy machine gun, firing from positions in the rear of the attacking troops, aims at centers of resistance within the hostile position, and prepares to give covering fire against counterattacks. Heavy machine guns follow the attacking rifle companies from position to position. In as much as unified control of this type of work is difficult, sections and platoons are usually placed under the command of the rifle companies to exploit local successes. Single guns may even be used in support of rifle squads or platoons to consolidate ground gained and to cover the flanks; however, this is practically the only time when heavy machine guns are used singly.

c. Defense

In defense the heavy machine gun is normally sited under the direction of the company commander. Sections may be placed under commanders of advance positions, or often, commanders of combat outposts. Otherwise, the heavy machine guns, although employed in the sectors of rifle companies, will form part of the battalion-fire plan. Their tactics involve exploiting all the possibilities of fire as early, as heavily, and at as long a range as possible. For this purpose positions are taken up in, or just to the rear of, the main line of resistance. Some heavy machine guns may be sited forward as "silent" guns. It is a German principle to site them for enfilade and crossfire. The heavy machine guns are sited in covered positions, with open positions forward of the main line of resistance having been reconnoitered beforehand and echeloned in depth. Thus the guns are able to move forward and engage any hostile force attempting to penetrate.

German doctrine stipulates that heavy machine guns in the rear of the main line of resistance may not fire overhead at ranges of less than 400 yards.

d. Conclusion

After the Battle of France a "Commander of Supporting Weapons" was instituted in the infantry battalion (normally the commanding officer of the machine-gun company). However, there is now reason to believe that heavy machine guns are increasingly being allotted in platoon strength (four guns) to the rifle companies, in whose sectors they are almost always employed in attack or defense.

German training stresses cooperation between the heavy machine guns and the battalion support weapons (infantry guns and antitank guns and mortars). The utmost attention is paid to the careful siting of fire positions, as well as of alternate and dummy positions. It has been found that the Germans make every effort to gain surprise, and that their camouflage is usually excellent.

Intelligence Bulletin Vol II, No 10. (June 1944), pages 6-9

Section II. GROUND TACTICS OF GERMAN PARATROOPERS

The commander of a German parachute demonstration battalion recently issued to his companies a directive which affords useful insight into some of the ground tactics that enemy paratroopers may be expected to employ. The following extracts from the battalion commander's order are considered especially significant:

1. For parachute and air-landing operations, I have given orders for section leaders and their seconds-in-command to carry rifles and for the No. 3 men on the light machine guns to carry machine carbines. There are tactical reasons for this decision. The section commander must be able to point out targets to his section by means of single tracer rounds. The No. 3 man on the light machine gun must be able to give this gun covering fire from his machine carbine in the event that close combat takes place immediately after landing. This last should be regarded as a distinct possibility. He must provide this covering until the light machine gun is in position and ready to fire. Before the assault, the No. 3 man on the light machine gun must also be able to beat off local counterattacks with his machine carbine until the machine gun is ready to go into action.

2. Since so many targets are likely to be seen only for a fleeting moment, and since the rifleman himself must disappear from hostile observation as soon as he has revealed his position by firing, the German paratrooper must be extremely skillful at "snap shooting" (rapid aiming and firing). The following three points are to be noted and put into practice:

a. Snap shooting is most useful at short ranges. It will not be employed at ranges of more than 330 yards, except in close combat and defense, when it will generally be employed at ranges under 1,100 yards.

b. Even more important than rapid aiming and firing is rapid disappearance after firing, no matter what the range may be.

c. Movement is revealing, also. Men must move as little as possible and must quickly find cover from fire at each bound.

3. I leave to my company commanders the distribution of automatic and sniper rifles within companies. I wish only to stress the following principles:

a. Wherever possible, sniper and automatic rifles will be given to those paratroopers who can use them most effectively. In general practice, this rules out commanders and headquarters personnel (who have duties other than firing).

b. There seems to be a general but incorrect impression that our sniper rifles improve the marksmanship of men who are only moderately good shots. These rifles are provided with telescopes only to make more distinct those targets, which are not clearly visible to the naked eye. This means that an advantage accrues solely to very good marksmanship firing at medium ranges -- and, what is more, only where impact can be observed and the necessary adjustments made. Since the sniper is seldom in a position where he can observe for himself, a second man, with binoculars, generally will be detailed to work with the sniper.

4. I wish commanders to make a report on the Battle of Crete on the subject of continual reference in their lectures, and in the lectures of platoon commanders who are training noncoms. I particularly desire that those passages in the report which deal with the importance of the undertaking as a whole be drilled into every man. The last three exercises I have attended have shown me that this principle is by no means evident to all platoon commanders. Platoon commanders in this battalion are still too much inclined to fight their own private brands of war instead of paying attention to the larger picture.

5. It is extremely likely that, during a parachute or airlanding operation, this battalion will land in hostile positions not previously reconnoitered, and will have to fight for the landing area. Such fighting will be carried out according to the same regulations which would obtain if we had fought our way into the heart of a hostile position.

6. Inasmuch as we shall soon be receiving our new machine guns[2] training with those new machine guns we already have must be pushed forward in our light companies -- at least to the extent of giving the No. 1 men about 1 1/2-hours a day on it. the most important point to be driven home is that this weapon is to be fired in very short bursts to avoid waste of ammunition.[3]

7. During the exercises and field firing demonstrations I have witnessed -- I admit they have been few -- I did not once see yellow identification panels used to mark our forward line, nor did I see the swastika flags used to identify our own troops to friendly aircraft. Henceforth, these panels and flags will be carried on all occasions and will be spread out at the proper times.

8. I wish platoon exercises to include more emphasis on the attacks on well prepared defensive positions. This will include cooperation between two assault detachments and a reserve assault ("mopping-up") detachment.

Each German paratrooper company commander, it is reported, must designate five to seven of his best men as a tank-hunting detachment. These men perform their regular duties, but are prepared to act as a team in their tank-hunting capacity whenever they may

---

[2] : Here the German battalion commander is probably referring to a consignment of regular or modified M.G. 42's.

[3] No doubt the enemy still hopes that this precaution will help to keep the fire on the target.

be called upon. The infantry training of German paratroopers is usually very thorough, covering all normal training, and, in some instances, use of the light machine gun, heavy machine gun, mortar, and antitank rifle, as well . Cunning and initiative are stressed. Many men are taught to drive tanks and other vehicles. Use of simple demolitions and the handling of antitank and antipersonnel mines are often included in the training.

Intelligence Bulletin Vol III, No 2. (October 1944), pages 7-11

GERMAN PATROL METHODS IN ITALY

The Germans are not interested in patrolling for the mere sake of establishing patrol superiority in no man's land; instead, they carefully regulate the extent of their patrolling according to the amount of information that they require. In recent fighting in Italy, German uncertainty as to Allied dispositions, together with the knowledge that Allied attacks were certain to take place, has led to a marked increase in patrol activity.

As to the regularity of German patrols, it can only be said that the procedure varies considerably, depending on the enemy units concerned. Some units sent out one or more patrols every night; others send out patrols only twice a week. The degree of activity is likely to be in inverse proportion to the amount of information that the Germans have already succeeded in collecting in the sector concerned.

TYPES OF PATROLS

Reconnaissance patrols, consisting of a noncom and three or four men, are sent out for the purpose of locating Allied positions and mine fields, and of noting any movement which might be significant. These patrols generally avoid contact, and retire hastily when fired upon. Very seldom is such a patrol given the mission of bringing back a prisoner of war.

Combat patrols consist of a noncom and eight men, at the very least, and usually are much stronger. In general 15 to 20 men are assigned to a combat patrol, with a sergeant in command. The patrol is likely to be organized in two sections of equal strength, each section having its own leader. A combat patrol customarily has the mission of bringing back prisoners of war. Because of the growing deficiency of German tactical reconnaissance and air-photo intelligence, prisoners of war have assumed an increased importance in the eyes of German intelligence officers.

Special patrols, whose strength varies according to the mission, are dispatched to perform demolitions, to engage Allied patrols who have managed to effect fairly deep penetrations and to prepare ambushes for supply columns.

Listening patrols, about the size of reconnaissance patrols or a little smaller, are sent out to discover whether soldiers in Allied positions are speaking English, French or Polish, or some other language which might have an order-of-battle significance, and to overhear any indiscreet remarks, which might be made by the men in the positions.

RECRUITING OF PATROLS

The normal enemy method of recruiting a patrol seems to be for a commander to order that a company provide a patrol of a certain strength. The company commander, in turn, sets a quota for each platoon. For a combat patrol, however, an entire platoon may be selected intact. Reconnaissance patrols probably are drawn from a single platoon. The soldiers who have been detailed for patrol work then report to company headquarters, where they are briefed by the noncom who is to lead them.

Even prior to the organization of special assault companies and assault platoons, an idea with which the Germans began experimenting several months ago, some units evidently made it a practice to always draw their patrols from a specific company. Either the same platoon was sent out over and over again, or a detachment of the "cross-section" type was sent out -- always under the same officer. In this manner, selected personnel became increasingly experienced in patrol work. It is believed that the enemy's use of specialized personnel work will be undertaken on a large scale in the future.

PATROL METHODS

As a rule, German patrols appear to depart at about 2200, and are not always given a time limit for their return. According to one source, a combat patrol moves in two sections, about 150 feet apart, and in this order: patrol leader, the two section leaders, a runner, and finally the two sections. Each man keeps a distance of about 15 feet between himself and the man in front of him.

Patrols are armed with a high proportion of light automatic weapons, always including at least one or two light machine guns, which are usually to cover the approach to any locality, or a withdrawal from it. The men usually keep their equipment down to a minimum, and frequently dispense with steel helmets. In snow-covered terrain, the Germans patrol in white uniforms.

Engineers often are attached to a patrol to guide it through German minefields and to clear a way through Allied wire and mines.

Also, artillery support may be given, in the form of harassing fire delivered shortly before the patrol reaches its objective.

Artillery fire may be delivered solely to mislead, and the patrol may go to an entirely different locality. Another deception involves artillery registration during the day, and artillery and mortar fire on the positions to be attacked during the night. As soon as the bombardment is lifted, the patrol places diversionary machine-gun fire on a flanking company and then attacks the positions which are the main targets.

German patrols may be sent out to test the strength of Allied outposts. If an outpost proves to be held weakly, the patrol attacks and seizes it, and remains there until relieved by troops from the battalion. In the event that the outpost proves to be garrisoned in some strength, the patrol tries to return with a prisoner of war, at least. It has been reported from a combat area on the Adriatic coast that specially trained dogs are used to approach buildings and give warning if they are occupied. However, the use of dogs for this purpose is not believed to be general.

Ruses are employed in an effort to trap Allied soldiers. Sometimes, when an Allied patrol or detachment approaches, only one German soldier will show himself, hoping to attract the force toward an ambush prepared by the rest of the German patrol, who opens fire simultaneously at a given signal. The Germans have been known to employ a similar trick when they wish to draw Allied troops within range of a covering machine gun; after the sentry in a position has given warning of the Allied approach, a decoy shows himself.

When a German patrol is challenged, it may try to pass itself off as an Allied unit by answering "Indian troops" or by mentioning some other Allied nationality, hoping that this will justify the foreign accent and thereby allay suspicion.

Imitation bird calls have been employed for signaling. One mechanical device in particular is said to emit a chirping sound in three tones.

After a raid in the Orsogna sector, a German patrol left behind a *Gebirgsjäger's* peaked cap complete with the edelweiss emblem. No Alpine troops were in the area, so the cap presumably was left behind to serve as a deception regarding the German order of battle.

If an Allied patrol is sighted, the German patrol may dispatch a runner to platoon headquarters to give information regarding the location, strength, and direction of the Allied force.

QUALITY OF PATROLS

The quality of German patrols varies considerably. For the most part, they seem to have been well trained, but at times they have given an impression of having been badly trained or else of being amazingly careless. Perhaps the Germans who display such a lack of security sense as to move quite openly along paths and roads are men who have not been in combat before. The uneven quality of German soldiers recently drafted is a factor to take into account; it is certainly true that a number of German soldiers have seized the opportunity offered by patrol work to slip through the Allied lines and surrender. In organizing assault companies, no doubt the Germans are motivated to a considerable extent by their anxiety to improve the quality of the patrols sent out.

Intelligence Bulletin Vol III, No 2. (October 1944), pages 12-18.

GERMAN FORWARD POSITIONS IN THE DEFENSE

German forward defensive positions, despite the widely varying terrain of different sectors are prepared according to certain standard German principals. This is the conclusion drawn from information supplied by numerous German prisoners captured in Italy.

FRONTAGES

No standard figures have been quoted for company or platoon sectors. However, forward weapon positions, each holding two men, are distributed fairly evenly, with distances of from 20 to 75 yards between positions, and with approximately four to eight men holding every 100 yards of front. Squad fronts, it is believed, are generally not less than 100 yards or greater than 250 yards.

DISPOSITIONS

There does not appear to be any great depth in the dispositions of forward companies. As German prisoners express it, these companies "put everything they have into the shop window." Battalions of four companies are likely to have three companies forward and one in reserve, although in at least one instance a battalion is believed to have placed all its personnel in the front line.

Positions consist of a series of foxholes or weapons pits, each holding two or three men. These positions usually are arranged in a single line. It is believed that the positions of the 2nd Platoon of the 2nd Company of a German reconnaissance battalion in very hilly terrain are worth considering, as a typical example of enemy procedure (see fig. 1).

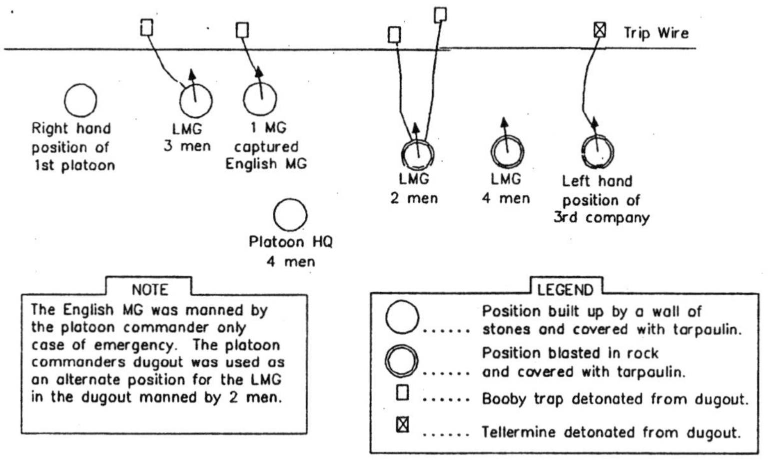

Figure 1.

In another instance, weapon pits were staggered in two lines 15 yards apart. The rear pits covered the intervals between the forward pits. The light machine guns were in the front line, one to each weapon pit. The riflemen were in the second line.

A short distance behind the forward foxholes or weapon pits, dugouts (*Wohnbunker)* were constructed, sometimes on a scale of one to each weapon pit. As a rule, these bunkers are connected to the weapon pits by communication trenches, but communication trenches between weapon pits are seldom prepared.

There is no new information regarding alternate positions.

It is said that, if listening posts are to be used, they are established close behind a wire-obstacle belt, and generally are manned by at least two soldiers.

WIRE AND MINES

All forward positions are protected by wire obstacles, mines, and trip wires. As a rule, wire is laid 50 yards or more to the front of positions. Trip wires are laid at about the same distance, and usually are one foot above the ground. Often such alarm devices as flare cartridges, tin cans, or bells are used in connection with wire.

There has been an increased use of mines, either in the form of minefields prepared a short distance in front of the positions, or single mines, some of which are detonated by wires pulled by the men in the positions when they detect a patrol approaching. Tellermines with pull-igniters, charges of as much as 40 pounds of TNT, and bundles of two or more stick grenades have been used for this purpose. Some weapons pits control as many as two or three such mines.

A prisoner remarks that, if the wire indicating a mine field is placed on the outer side of the surrounding states, there is a genuine minefield within the wire; on the other hand, he declares, if the wire is placed inside the stakes, it is serving only as a trip wire. Pending further evidence on this point, the prisoner's statement should be treated with reserve. Even so, it is worth noting as possibly having been true in certain instances, and as a method which may be repeated in the future.

INFANTRY SUPPORT WEAPONS

There is little new information regarding the siting and allotment of infantry support weapons, but two German prisoners mention the allotment of additional machine guns from company reserves to strengthen weak or supposedly threatened sectors. A prisoner from a heavy infantry-gun company reports that his company's guns were in "support readiness" 1.8 miles to the rear of the infantry positions. Several reports have referred to the complete lack of activity on the part of infantry guns. Mortars are sited on reverse slopes, wherever possible. Figure 2. illustrates a mortar position reported from the Italian Adriatic coast. The weapon pits were 6 feet in diameter and 4 feet deep. The ammunition pits and communication trenches were 4 feet deep. There were 35 bombs for each mortar, a figure above the average.

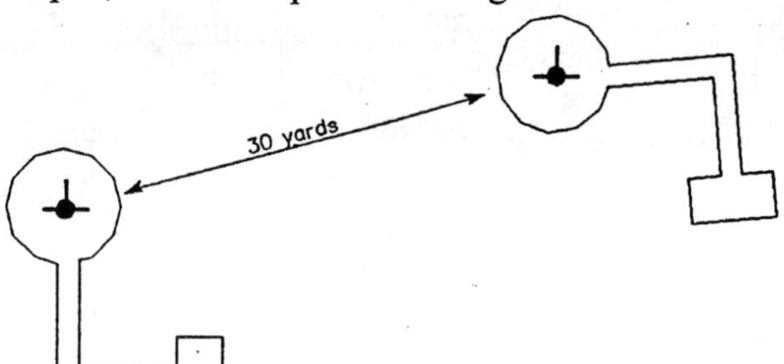

Figure 2.

COMPANY AND PLATOON HEADQUARTERS

Information on the subject of company and platoon headquarters is also scant. A prisoner states that his platoon's headquarters was 150 yards behind the middle squad (a distance which is believed to be more or less average) and that company headquarters was 1,600 yards to the rear (a distance which is believed to be unusually great.)

MANNING OF POSITIONS

The general German practice is to man weapon positions 100 percent at night, with two men to each position, and to have no more than 50 percent of personnel manning them during the day, with the men off duty sleeping in dugouts. Machine guns are manned day and night. In some units, only the machine guns and observation points are manned by day, the remaining personnel withdrawing to the dugouts. Movement in and around positions visible to Allied forces is reduced to a minimum, and experienced units undertake virtually no movement at all. Newcomers to the front usually may be identified by their initial clumsiness in this respect.

AMMUNITION SUPPLY

Figures for the number of rounds of the different types of ammunition vary greatly. For light machine guns and heavy machine guns, the supply usually will not be less than 1,100 rounds for each machine gun, or more than 3,000 rounds. As a rule, the figure is somewhere between 1,250 and 2,000. Riflemen get 90 to 100 rounds (120 for semiautomatic rifles). Mortars are allotted 20 to 25 bombs; cup dischargers, 20 to 30 grenades; and machine pistols, 100 to 180 rounds. Tracer normally is used in a proportion of one to five, but one source states that the use of tracer in the MG42 is being abandoned since the high rate of fire produces an almost continuous flare. This flare is very easy to detect at dusk or in darkness, and betrays the position.

COUNTERATTACKS

Theoretically, companies are not supposed to commit more than 80 men at any one time, in order to maintain a reserve for counterattacks. It is known that counterattacks are included in the missions assigned to assault companies and platoons now being formed in that number of units. The German intention may be to relieve the ordinary rifle companies of much of the responsibility for counterattack, but the individual company or platoon will be obliged to make the first quick counterattack (*Gegenstoss*) if valuable time is not to be lost while the Allied force consolidates.

LIGHT SIGNALS

Very light signals are used in all sectors for recognition purposes, to call for artillery support, and to give warnings. The procedure varies somewhat, depending on the units involved; some units change signals nightly, while others change them at irregular intervals or almost never. Much seems to depend on the individual commander. However, it has been observed that the better units change signals frequently.

Signals appear to fall into two classes: those used during mobile operations and those used when conditions are static.

It is said that signals used during mobile operations are not likely to be altered, because of the difficulty of keeping all elements informed about the changes. Single lights are used for such signals, and the following code nearly always is employed:

White	"We are here."
Red	"Enemy is attacking."
Green	"Request artillery fire."
	or
	"Advance the artillery barrage."
Blue or Violet	"Enemy tanks."

Light signals under static conditions are reported to be combinations of white, red, and green, as a rule, with no more than two colors being used at a time. White still appears to be used chiefly for recognition of German positions. Combinations of red and green thus far have been used only for signals requiring action. Such signals may lead one to expect German activity of some kind.

To give an alarm, forward positions generally use whistling cartridges.

Intelligence Bulletin Vol III, No 2. (November 1944), pages 89-93.

WITHDRAWAL TACTICS IN ITALY

PATROLS

From time to time, U.S. enlisted men who are fighting the Germans in Italy provide the *Intelligence Bulletin* with helpful first-hand information regarding tactics and weapons of the enemy's small units and of the individual German soldier. This month a number of noncoms who have been in action in various combat areas of the Italian theater furnish us with some fresh information about a enemy combat methods and also corroborate information previously reported.

——

Patrols

"The Germans sometimes use suicide squads of about 50 men to determine the strength of an Allied position. I have observed this procedure on several occasions. One instance is particularly worth noting. Our patrol was holding a hilltop. The Germans knew the approximate location, but did not know our strength or fire power. The enemy sent a patrol of about 50 men to determine our strength. One of four outposts succeeded in spotting the German soldiers while they were some distance away. From the Germans' behavior, the outpost deduced that an officer was giving them instructions. Since their advance was anticipated, we were ready for them when they approached. The 50 men were so deployed that the distance from the left flank to the right flank measured no more than 75 yards. The Germans approached through the darkness, making no very great effort to maintain silence. As a result, it was possible to hear them while they were still some distance away. We allowed them to reach a point about 30 or 40 yards from our line, when we opened up with everything we had. The Germans, who were directly in front of my company, returned fire. Since they lacked good cover, about 20 or 25 were killed. Soon the others had to seek cover. The balance of the patrol did not surrender until they had been brought out of hiding by white-phosphorus grenades."

——

Engineer Ruses

"While we were clearing mines from a beach, I found a number of Tellermines embedded in the sand. About 6 of these had been laid fairly close together, and when these 6 were detonated, a field of about 25 or 30 mines exploded simultaneously. The entire group had been connected with primacord to produce this result. As our work progressed, we found that the Germans had gone in extensively for this method of booby-trapping."

"On one occasion, at the Rapido River, my company had cleared a road through a minefield and had marked the clearing with tapes. Our unit moved ahead without leaving a guard at the clearing. During the night the Germans infiltrated into the area. They discovered the tapes, moved them some distance from the clearing, and set them up again, marking a road directly through a minefield. The Germans were well aware that an infantry advance the following morning would have to be delayed until engineers could clear another pathway for the advance."

Machine-gun Tactics

"The Germans are strong believers in automatic weapons, and use them in greater numbers than we do. The Germans seem to operate on the theory that if they fire enough ammunition into an area occupied by our troops, they are bound to inflict enough casualties to make the expenditure profitable, whether or not they aim."

"Enemy machine-gun positions sometimes consisted of two concealed guns sited to provide converging fire and a third, less well concealed, off to one side. An observer is posted between the two concealed machine guns. The third machine gun is fired in a deliberate effort to attract attention, and tracer bullets are used. The other two guns remain silent. This ruse is intended to deceive Allied troops into thinking that they face only one machine gun and that they have spotted this position accurately enough to warrant an attempt to take it. The two concealed machine guns await just such an attempt, with the hope of causing many casualties by surprise fire."

"The Germans experience considerable difficulty in holding down their machine pistols and machine guns. Generally speaking, the first few rounds of a burst are at the desired height, and the remainder are too high."

"Often enemy machine guns are sited in groups of three across a suitable route of advance for an Allied force. The intention is to permit an advancing Allied squad to reach a point about even with the flanking guns, which are forward of the center gun, the latter forming the apex of a triangle. A flanking gun opens up, and when the Allied soldiers direct their attention to the gun, the one dead ahead starts firing. Usually the third gun will wait until the Allied soldiers have moved far enough forward to permit it to place fire on them from the rear. The German idea is that even after our men succeed in silencing one or two of the guns, the third still may make the Allied advance a costly one."

Combat in Hilly Terrain

"In hilly terrain the Germans make special efforts to keep tabs on the movement of Allied forces. A system of listening posts seems to be maintained. Using tracer bullets, the Germans fire machine pistols towards the U.S. troops moving in the hills. The German mortars immediately begin shelling the posts indicated by tracer."

"The German combat patrols I've encountered have consisted of from 9 to 16 men armed with machine pistols, light machine guns, and light mortars. I've noticed that these patrols fan out and allow only one man to fire at any one time, so as not to disclose the whereabouts of the rest of the patrol. Also, I've observed that German patrols don't skimp on the quantity of ammunition they carry."

"German outposts use colored flare signals to call for mortar or artillery fire. These colors don't always mean the same thing -- that is, in the areas where I've been in an action. German artillery and mortars are able to 'zero in' on advancing forces almost immediately after the signal flares have been sent up. As soon as we found this out, we were prepared to meet an artillery or mortar barrage whenever we saw flares of a significant color or colors."

———

More on Target Designation

"The Germans use their machine pistols as a means of communication. For example, I've observed that a German patrol might fire two or three bursts, which would be answered a few seconds later by other machine pistols in different areas along the front or behind it. Later the nearest German patrol would fire tracer bullets over a certain area, and then, after approximately 4 or 5 minutes, the Germans would place heavy mortar fire on the area indicated by the tracer. It was the consensus of opinion among the men in our platoon that a German patrol would spot a target and then communicate with German mortars and artillery, pointing out the exact location of the target."

———

Cover Trap

"Thirteen other men and I had just reached a prepared ditch, which the Germans had deserted. We were about to start digging foxholes in the ditch when German 88's suddenly began firing on us. I am of the opinion that the Germans deliberately prepared these ditches with the intention of luring U.S. troops into using them and of placing registered fire on us as soon as we did so. The 88's were fired in bursts of five, always followed by a lull of about 10 seconds. This lull was general, and would seen to indicate that all guns were loaded and fired simultaneously. It was during one of these lulls that I managed to escape from the ditch."

Intelligence Bulletin Vol III, No 2. (November 1944), pages 86-89.

TACTICS OF PERSONNEL CARRIERS MOUNTING FLAME THROWERS

The official German Army directive which is paraphrased below discusses the tactics of the armored half-track flamethrower vehicle (see figure 21). This is the medium personnel carrier used in the Panzer Grenadier units, fitted with two large flame throwers and one small flame thrower, in addition to the regular machine-gun mounts at the front and rear.

A large flame thrower is mounted in a V-shaped shield on each side of the vehicle. The small flame thrower on the back of the vehicle is simply the cartridge ignition projector used in the small portable flame thrower; this projector is attached at one end of a 33-foot length of hose, which connects to the propulsion unit and fuel tank. Tanks situated next to the side armor plates, in the interior of the vehicle, carry 154 gallons of flame-thrower fuel. A small gasoline engine and a pump, used to propel the fuel, are situated in the center of the interior.

The effective range of the large flame throwers is about 40 yards, whereas that of the small equipment is unlikely to exceed 30 yards. The fuel carried is sufficient for about 80 bursts, each lasting from 1 to 2 seconds.

---

1. The medium armored flame-thrower vehicle is a close combat weapon of the Panzer Grenadiers. It is used in the offensive when the other weapons used from the vehicle do not promise to be sufficiently effective.

2. In addition to employing its machine guns against personnel at rages of as much as 440 yards, the vehicle may direct flame against personnel and static targets at ranges of as much as 40 yards. If the flame does not destroy hostile troops, it should at least force them to leave their cover. Attacks with flame are particularly effective in mopping up ground quickly, in liquidating hostile soldiers who put in a sudden appearance near the vehicle, and in destroying hostile personnel in hasty permanent field fortifications.

Figure 21. Armored Half-track Flame-thrower Vehicle.

3. The flame-thrower vehicles normally are employed by whole positions, and always in close cooperation with mounted Panzer Grenadier units in the attack.

4. For combat in fortified areas, attacks on permanent fortifications, and so on, the vehicles may be employed singly, under the command of mounted Panzer Grenadier platoons. When the latter dismount, the flame-thrower vehicles will be left behind with the armored personnel carriers.

5. It is forbidden to use these vehicles like infantry tanks or assault guns, as "point" vehicles on the march or in action, for protective duty, or as independent patrol vehicles.

6. In the pursuit, the platoon will support local and prepared counterattacks by mounted Panzer Grenadiers.

7. Every effort must be made to employ the platoon as a whole, for greater effectiveness.

TACTICS

1. Preparations for the attack (terrain estimate, tactical reconnaissance, protective duties, camouflage, and so on) follow the same principles are as observed by tanks and panzer grenadiers.

2. In the attack the flame-thrower vehicles move in extended order behind the mounted panzer grenadier units. The action normally is opened by machine-gun fire. Covered by the fire of other weapons, as well as by the weapons in the personnel carriers themselves, the flame-thrower platoon will break into the hostile position.

3. If the opposition remains under cover, it will be burnt out. Bursts of fire from the flame throwers should be projected only against those targets which definitely are within range. To fire flame bursts indiscriminately, before reaching the opposition, merely wastes fuel and obscures vision.

4. It is important to direct the flame against the bottom of the target first and then work up, so that hostile personnel who may have close-range antitank weapons in readiness will be destroyed.

5. The type of target and the course of attack will determine whether fire is to be opened while on the move or at the halt.

6. Trenches will be crossed and engaged from the flank. Tree tops, roofs, and raised platforms may be set afire if the presence of hostile soldiers is suspected.

7. If a large conflagration is desired, the target first will be sprayed with oil and then ignited by a burst of fire. This is especially effective when attacking dugouts, trenches, entrances to pillboxes, and -- of course -- wooden buildings.

8. Fire will not be opened in thick, natural fog, except by special order.

Intelligence Bulletin Vol II , No 2. (October 1943), pages 21-23.

Section V. STREET FIGHTING BY PANZER GRENADIERS

German Panzer Grenadiers (armored infantry) are given extensive training in street fighting. Cooperating with tank units, the Panzer Grenadiers are often employed for the close-in combat that is required when the Germans wish to put an end to all resistance within a town -- generally one which has been, or is being, encircled. The following extracts are from a Panzer Grenadier lieutenant's account of such an action. In spite of its Nazi point of view and its heightened style, it is interesting as an illustration of Panzer Grenadier activity. The action the lieutenant describes takes place on the Eastern Front. A well-deployed tank battalion, followed by a Panzer Grenadier company riding in armored personnel carriers, is advancing across the plains of the Ukraine.

A Russian force has been encircled, and the task for today is to drive through the center of the pocket and divide the Russians into still smaller groups, which can be destroyed separately. As yet, no rounds have been fired, but the tanks ahead of us may come upon the hostile force at any moment. The company commander glances at his platoons; they are following in considerable depth and width. The distance between vehicles is at least 50 feet, the radios are set for reception, and everything is in order. It is very hot and there is a haze.

The men in the tanks ahead can see a village in the distance. According to the map, this should be Krutojarka. Guns can be seen flashing at the edge of the village. The Russian force is engaged. We hear the fire of the Russian antitank guns and our own tank cannon, and, in between, the sound of both sides' machine-gun fire. The Panzer Grenadier company commander gives his command by radio. As soon as the grenadiers see Russian soldiers, they are to fire on them directly from the personnel carriers, or else dismount quickly and fight on the ground, depending on the requirements of the moment.

The first tanks enter Krutojarka, but presently reappear. The company commander gives the radio command, "Krutojarka is being held by the enemy. Clear the town!" The personnel carriers advance past the tanks, which are firing with all their guns, and move toward the edge of the village.

A personnel carrier's tread is hit by a flanking antitank gun. The grenadiers jump out and assault the antitank- gun crew with machine-gun fire, while the driver and the man beside him get out and, under fire, change the broken link of the tread.

The attacking grenadiers have now reached a street at the edge of the village. Startled by the suddenness of the assault, the Russians take cover in houses, bunkers, foxholes, and other hideouts. The grenadier jump out of the personnel carriers and advance along the street, making good use of grenades, pistols, and bayonets. The driver and a second man remain in each carrier.

The personnel carriers skirt around the sides of the village, with the men beside the drivers delivering flanking fire against the buildings. Soon the roofs of the houses are afire. The smoke grows thicker and thicker.

Three tanks push forward along the main street of the village to support the attack of the grenadiers. We find the smoke an advantage, for it prevents the Russians from discovering that there are relatively few of us. Also, as a result of the poor visibility, the Russians cannot employ their numerous machine guns with full effect. We, for our part, are able to engage in the close-in fighting at which we excel. It is no longer possible to have one command for the company. Officers and noncoms have formed small shock detachments, which advance from street corner to street corner, and from bunker to ditch, eliminating one Russian nest after another.

A lieutenant holds a grenade until it almost explodes in his hands, and then throws it into a bunker. It explodes in the firing hatch, and enemy soldiers stream out.

The company commander discovers a 37mm Russian antiaircraft machine gun, and sits down on the saddle. Two men who are with him attach the magazines, which are lying about. Although the commander has never fired this type of cannon before, he succeeds in demoralizing the Russians with its high-explosive projectiles. We take many more prisoners.

When about half the village is in our hands, and when we have captured the Russian commander and his political commissar[4], resistance collapses. All prisoners are marched to the rear, and the booty of guns and vehicles is collected. The Panzer Grenadiers advance to the far end of the village, where they climb into the waiting personnel carriers. Most of the tank battalion also has skirted the village, and already has moved further east. Anticipating further action, the Panzer Grenadiers again follow the tanks.

Intelligence Bulletin Vol II, No 5. (January 1943), pages 41-46.

Section VI. GERMAN COMBAT TACTICS IN TOWNS AND CITIES

1. INTRODUCTION

Now that United Nations forces are fighting energetically on the soil of continental Europe, it must be expected that we shall engage the enemy in towns and cities with ever-increasing frequency. For this reason, it is most important for us to understand German doctrine regarding combat in populated places.

Often the size of a town is not the principal determining factor in a German commander's tactical plan; instead, a town's geographic or economic importance may be his first consideration. A very small village may be worth defending fiercly if it commands the entrance to a mountain pass, for example or if it posses resources essential to the German war effort. A much larger town, on the other hand, may have far less value, and by no means be worth the same expenditure of men and matérial.

However, the Germans use the same basic tactics for towns and cities alike. These tactics are summarized in the following paragraphs, which are based on a German Army document.

---

[4] Since this account was written, the Russians have discontinued their practice of assigning political commissars to accompany Red Army units.

2. IN THE ATTACK

The Germans attempt to outflank and encircle a town. If this move succeeds, they cut off the water, electricity, and gas supply. They find the most vulnerable spot in the area held by the hostile forces, and penetrate it. After cutting the hostile forces in two, the Germans then divide the opposition again and again so that it no longer is able to maneuver freely. German doctrine maintains that parallel attack constitutes the most advantageous method if a number of columns are available (see fig. 17a). Thrusts at an angle (see fig. 17b), and especially thrusts from opposite directions (see fig. 17c and d), are avoided. The Germans believe that such thrusts are likely to result in friendly troops getting under each other's fire, and that confusion is inevitable.

The Germans group their units in attack columns and mopping-up columns. Advance by limited sectors is the rule. Commanders do not plan too far ahead. After taking a block, a commander reassembles his men and issues further instructions.

It is a German axiom that "he who commands the heights also commands the depths." The Germans drive for dominating positions, although in defense they may site their machine guns in much lower positions.

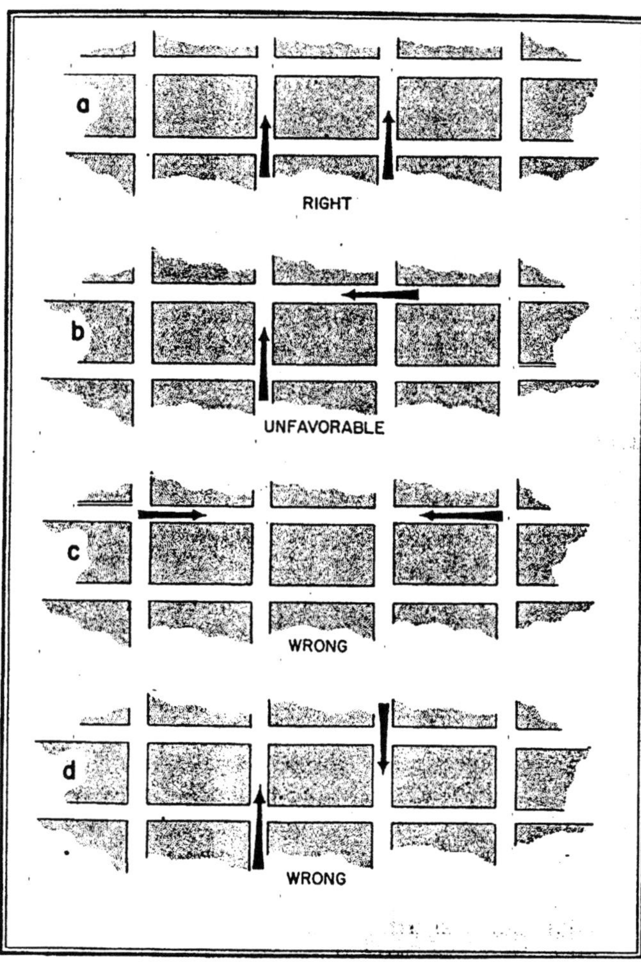

While an advance is being made in a street, a simultaneous advance is made along the roofs if the houses adjoin each other. In the streets an advance is made by single files edging forward close to the houses on each side. From front to rear, marksmen are detailed to observe danger points on the opposite side of the street; that is, a man may be ordered to observe all roofs or all the windows on a given floor, and so on.

Side alleys and entrances to side streets are blocked as rapidly as possible. After this, searching parties are detailed to investigate all buildings in the block and then close them. Entrances to cellars and first floors are guarded until all houses in the block have been searched. All windows are closed and shuttered.

Figure 17. - German Attack Tactics in Towns and Cities.

The Germans attach importance to constant and effective cooperation with artillery. Light guns (manhandled) advance along the street and combat nests of resistance with direct fire. The Germans believe that aside from air bombardment, only 150mm and 210mm pieces are effective in destroying the larger buildings. Tanks, they maintain, are not very successful in assaults on houses.

When attacking a town, the Germans do not employ motor vehicles or horses. As much gear as possible is sacrificed in favor of axes, crowbars, wire-cutters, saws, ropes and rope ladders, flashlights, prepared charges, hand grenades, smoke candles and grenades, and maps and air photography of the locality.

3. IN DEFENSE

In organizing the defense of a town, the Germans prepare a reserve of drinking water, rations, ammunition, and medical supplies. Some of this is stored in various cellars, since the Germans are aware that these are tactically useful places, from which the advance of a hostile force can be hindered considerably. Searchlights are kept ready to illuminate the target area at night. Preparation is made for defense by sectors. Mines and booby traps are kept ready for use.

So that the main line of resistance will not be discovered by the hostile forces, the Germans place it within the town proper and make it irregular. Only individual centers of resistance are established in the outskirts. These are used for flanking purposes. Important buildings are defended, not from their own walls, but from advance positions.

The Germans try to maintain a strong mobile reserve.

Every attempt is made to trap hostile units in dead-end streets and to cut off, by sudden flanking movements, hostile units which advance too recklessly.

Emergency barriers are kept ready in the entrances to buildings. These barriers can be placed across the streets on extremely short notice.

Low machine-gun positions are prepared to cover all possible approach routes.

Good telephone communication is maintained.

Command posts remain constantly on guard against surprise attacks.

The windows of all buildings are kept open at all times, so that the attackers will have difficulty in detecting from which windows fire is being delivered. German soldiers are instructed not to fire from window sills, but from positions well within the rooms. Snipers continually move from room to room. Individual roof tiles are removed to provide loopholes. On rooftops, firing positions behind chimneys are considered desirable, provided that the chimneys are below the roof ridges.

Important doors leading to the street are guarded, and doors which are not to be used are blocked. Holes are pierced through the walls of adjoining houses to afford communication channels.

The Germans regard the entire matter of defense merely as a preliminary to surprise counterattacks.

Intelligence Bulletin Vol II, No 5. (January 1943), pages 20-24.

Section II. HOW INFANTRY BATTALIONS DEVELOP FOR THE ATTACK

1. INTRODUCTION

The German theory of entrance into offensive combat is fairly usual, in that two distinct stages are involved. These are called the *Entfaltung* and *Entwicklung*, which may best be translated into U.S. terminology as "development" and "deployment." The first stage is evidently designed to permit more rapid deployment at the proper time, and to enable good control to be maintained until as late a moment as possible. Briefly, the first stage (*Entfaltung*) begins with the approach march, when the battalion changes from a route-march formation to one made up of several columns. The second stage (*Entwicklung*) covers what, in U.S. practice, is the deployment of platoons and squads. The following paragraphs outline the tactics involved in each stage, as they are taught to German infantry noncoms.

2. FIRST STAGE

Normally, the development of a regiment is by battalions (see fig. 10). If necessary, distances between battalions are increased.

When a high state of preparedness is necessary, the battalion itself may "shake out" into companies. Companies proceed in the direction given them, employing the normal marching formation and, at the same time, making use of whatever cover and concealment are available. Commanders take into account the additional strain of marching across country.

Company transport remains with companies as long as possible, until the companies themselves must deploy.

The Germans believe that it is often advisable to have only one company forward, with the main strength kept directly under the battalion commander as long as possible, ready to be employed in the direction most advantageous for an attack.

Support weapons are used to cover the development and the subsequent advance. These weapons are interspersed in the line of march, either between the companies or behind the battalion. If any alteration of intervals is caused by ground conditions or hostile fire, the original intervals are resumed at the earliest opportunity.

When the development takes place, the leading elements of the battalion may be ordered to seize tactically important terrain.

German training points out that deployment at night, and in woods, calls for stronger protection forward for preparatory reconnaissance, and for the marking of routes. Intervals between the units are shorter than by day.

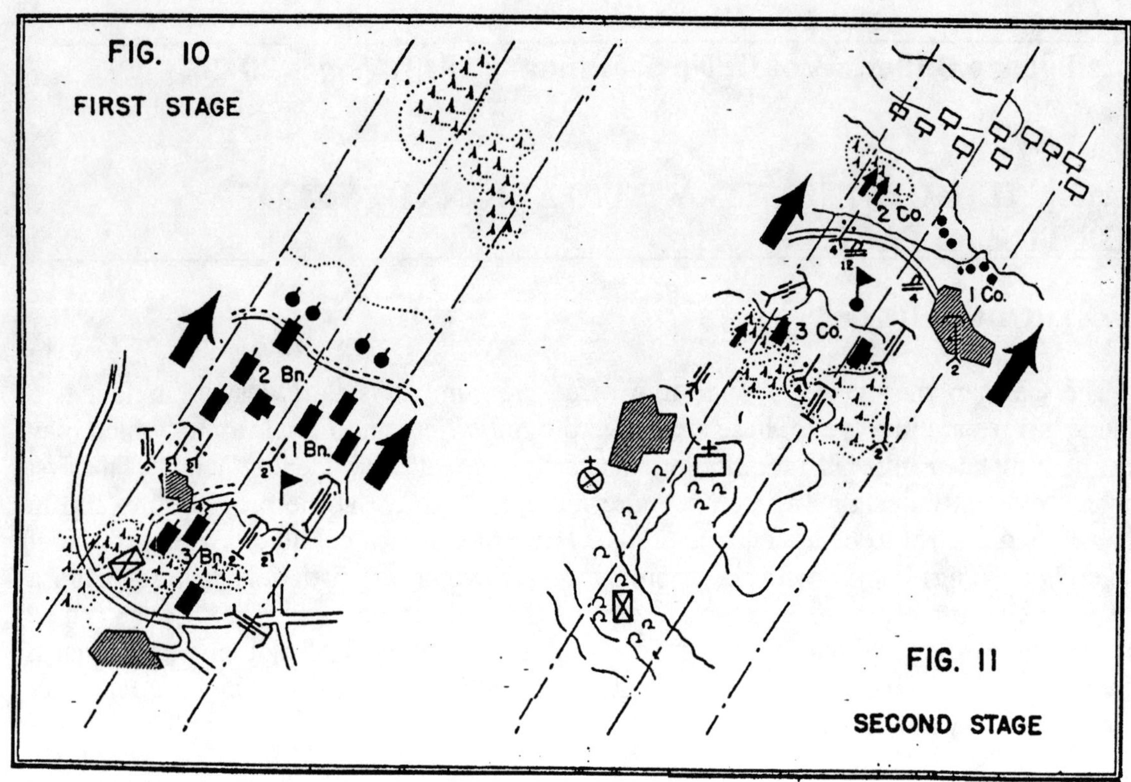

The battalion commander's orders cover:
    a. Information regarding hostile and German units;
    b. German intentions;
    c. Reconnaissance;
    d. Instructions for forward companies (including seizure of dominating terrain);
    e. Instructions for weapons supporting the advance;
    f. Instructions for the companies comprising the main body (including reconnaissance on flanks and protection of flanks, if necessary);
    g. Rendezvous of company transport and battalion vehicles;
    h. Battalion headquarters; intercommunications.

At the time of the development, the battalion commander moves with the forward elements from one prominent terrain feature to another. He generally sends special reconnaissance patrols ahead, or reconnoiters the hostile position himself from a commanding terrain feature. Commanders of support weapons accompany him, reconnoitering for firing positions.

3. SECOND STAGE

As soon as any German company comes within rage of observed hostile artillery fire, it disperses in depth (see fig. 11). The Germans consider that an advance in file is often desirable, inasmuch as it represents only a small target and one which is easily controlled; they recognize, however, that it is highly vulnerable to enfilade fire.

If ground conditions and hostile fire make deployment necessary, the platoons may be dispersed in depth into squads. The Germans find that this splitting-up permits the ground to be exploited for cover during the advance, and that it hinders the effectiveness of hostile observation and fire.   Reserves and support weapons also adopt open formations. They remain sufficiently far behind the forward elements to avoid coming under fire directed at these elements.

If the rifle companies are deployed, their elements normally move forward in narrow columns or single files, with irregular intervals, and make use of all available cover. The forward elements are not deployed as skirmishers until they are required to engage in a fire fight.

Intelligence Bulletin Vol II, No 6. (February 1944), pages 71-74.

Section IV. CLOSE-QUARTER FIGHTING AND WITHDRAWAL

The battle of Primsoloe, which took place during the Sicilian campaign, furnishes a very good example of German tactics in close-quarter fighting and withdrawal.

The initial assault by United Nations forces was made on the morning of 15 July 1943. After bringing it to a standstill, the Germans made no attempt to defend the river line, but concentrated on holding a position in the vineyards and ditches on each side of the road, north of the bridge. This position was based on a sunken trail, which ran west from the main road, about 200 yards north of the river, and which afforded concealment. Shallow trenches had been dug in the banks of the trail. The Germans also made use of ditches, which ran east and west from the main road. Pillboxes in that area had been engaged by 75mm gun fire from United Nations tanks, and for this reason were not used by the Germans.

The Germans were equipped with a very high proportion of automatic weapons, especially light machine guns. At night, light machine guns fired on fixed lines very close to the ground. The fire was coordinated with the firing of flares. Bursts of 10 to 15 rounds were fired at a rate of about one burst every minute.

In the daytime, German machine guns were well concealed in commanding positions in ditches and along the sunken trail. Extensive use evidently was made of alternate and supplementary positions, for each machine gun appeared to fire first from one spot and then from another. Never more than two, or possible three, machine guns were firing at any one time. This suggested the presence of a very small force, whereas the length of the sunken trail alone (from 200 to 300 yards) the number of rifles and other weapons subsequently counted, and the number of prisoners taken, indicated that there were at least 50 to 60 men.

Individual snipers armed with light machine guns, submachine guns, or rifles were concealed in the vineyards and trees forward of, and on the flanks of, the main German position. The mission of these snipers probably was to protect the German flanks and to harass the United Nations force.

During the first part of the battle, the Germans had very few mortars. Only one is known to have fired; its fire was inaccurate and evidently not observed, perhaps because of the closeness of the fighting.

Grenade-throwing pistols and rifle grenade dischargers were used at close quarters to put down a heavy concentration of high explosive. Both types of weapons throw a high-explosive grenade approximately 20mm in diameter. Many stick grenades and egg grenades also were used.

The Germans had four or five 88mm guns and one or two antitank guns of smaller caliber, 20mm or 37mm. These guns were used principally to cover the main road. No attempt was made to conceal them, probably because they were brought up in great haste when the Germans discovered the presence of United Nations tanks and realized that demolition of the bridge was impossible. However, individual Germans concealed themselves in ditches by the side of the road and in the culvert under the road, and engaged our tanks at close quarters with demolition charges and magnetic antitank grenades.

The German withdrawal from the defense position was accomplished at the rate of 5 to 6 miles daily. Each day the movement was made to a position previously selected. Commanding ground was the deciding factor in the choice of their positions, which afforded good fields of fire for machine guns and good observation posts for mortars. Sometimes the positions were based on natural antitank obstacles, such as riverbeds. Towns and villages were not used as centers of resistance, except where positions commanding a bottleneck could be obtained by the expedient of occupying houses situated on high ground. Once the guns occupied a line of houses built on a very high ridge. A sunken road behind the houses provided good lateral communications and a covered line of withdrawal.

Patrol reports and reports from civilians indicated that the Germans usually withdrew in the early morning, between 0200 and 0400 hours, the last elements to leave often being protected by a few tanks. The type of fire, which had marked German withdrawal in Africa -- increased shelling and machine-gun fire at the end of the day and at intervals during the night -- was not employed here.

Intelligence Bulletin Vol. III, No 8. (April 1945), pages 68-71

WHAT THE GERMANS LEARNED AT WARSAW

The Germans gained so much experience in street fighting while suppressing the first general uprising of armed Polish forces in Warsaw that the official German "Notes for Panzer Troops" takes cognizance of the lessons learned during these operations. Employing the popular German Army training method of listing incorrect and correct procedures in parallel columns, "Notes for Panzer Troops" sets forth a number of German errors in the Warsaw fighting, and supplies official comment on the methods which should have been employed in each case.

The observations in the "right" column take on an added significance, in view of a statement by the Inspector General of Panzer Troops to the effect that the underlying principles must be applied in *all* street fighting.

WRONG	RIGHT
A large number of heavy support weapons (assault guns, heavy howitzers, assault howitzers, self-propelled antitank guns, and infantry heavy howitzers) were used in an uncoordinated fashion, with a consequent lack of effect. The fire of the support weapons was not used for immediately pushing forward.	1. All available support weapons, including artillery and aircraft, are concentrated on approved targets. During the concentration the infantry prepares to attack as soon as the last shell has fallen. Armored vehicles accompany the infantry, to keep down any hostile soldiers who are still alive and who try to reappear.
Our troops mainly used streets. (In street fighting the enemy can take advantage of innumerable hiding places. Absence of visible enemy therefore by no means implies an absence of enemy.)	2. Walls of adjoining houses are blasted, and troops move forward through the houses. Mopping-up parties of infantry follow (making such covered approaches facilitates evacuation of wounded and supply of ammunition and rations.
Houses or blocks were not consolidated immediately after capture. Infantry lingered around the entrances, doing nothing.	3. As soon as a building has been taken, it is consolidated; windows and other openings are turned into firing ports. Since underground passages and sewers provide the enemy with cover and means of communication, the entrances to cellars, stairs, and so on are to be given special attention. If subterranean passages cannot be mopped up immediately, the entrances must be barricaded, or blown in and guarded. Troops will not stand around idly.
Completely ruined houses were regarded as being no longer of use to the enemy. (It was found, however, that the enemy made considerable use of completely destroyed buildings.	4. Even the most completely ruined houses must be occupied or covered by fire. Roving patrols are detailed to deny access to them and to ferret out any hostile stragglers who may have occupied them.
Many of the houses that we occupied had been almost completely destroyed by our own fire, thus denying our attacking troops positions and cover.	5. As far as possible, random destruction of potential cover can be prevented by strict discipline. Only outbuildings affording the enemy covered approaches to destroy vital points should be destroyed.
Armored vehicles were used to knock down barricades and walls, to push aside abandoned vehicles and guns, and to perform other tasks for which they are not suited.	6. The fire power of the armored vehicles must be conserved by all means. In street fighting they are very much exposed to close-range antitank weapons. This makes them fundamentally unsuited for "bulldozer" tasks. The accompanying infantry therefore protects them against surprise attack of any kind. When attacking barricades and obstacles, the infantry approaches first and forces a passage through the obstacles. Squads of civilians later are put to work to complete clearing of debris.
Our troops failed to make sufficient use of their rifles. The enemy was not sufficiently harassed.	7. Rifle and machine-gun fire must be delivered promptly and steadily from all newly captured buildings. Rifle fire is concentrated on group targets to keep the enemy's head down. The enemy is not given a moment's rest, but feels himself perpetually observed and engaged. Rapid opening of fire is especially important,

The supposedly non-combatant and "harmless" population was not kept under observation, and seldom was employed to clear debris.	to avoid giving the enemy time to withdraw to alternate positions. 8. All able-bodied civilians are employed to clear debris. The German Army must enforce this point relentlessly, even when work is performed under fire. (In this case the whole population was more or less directly assisting the insurgent Polish troops.)
Sufficient cunning was not employed to counter the enemy's tactics.	9. Tricks must be employed to draw fire and silence it. Our methods must change constantly; feints and other tricks and imaginative tactics must be devised.
The liaison between the various assault detachments was generally too loose, and signal communications were inadequately used. Radio and telephone conversations were practically always in the clear.	10. Assault detachments are instructed in methods of cooperation, use of fire, and movement. Cooperation will be improved if the assault detachments are kept constantly in picture and if they report regularly on their position and intentions.

The Inspector General of Panzer Troops adds a final comment of his own. He says, "When tanks are used in street fighting, they should be employed like the so-called 'tank infantry teams' used in Normandy -- that is, small infantry units will be detailed to cooperate directly with tanks. The tougher the fighting, the greater our casualties will be if the following principles are not observed: (1) no splitting of forces, (2) thorough and purposeful concentrations of fire, (3) immediate infantry exploitation of tank fire and (4) the closest mutual support throughout each action.

Intelligence Bulletin Vol III, No 10. (June 1945), pages 52-56.

TACTICS OF A GERMAN PATROL

GENERAL SITUATION

"Captain Stouffer's Troop 'C' was employed on a line in contact with the enemy (see accompanying map). Platoons occupied outpost and dugout positions with mutual fire support, but extremely poor observation. (Hedgerows and orchards were profuse.) Communications depended strictly upon line-wire telephones, with radios in reserve for emergency use."

"The main obstacle was a canal ('No Man's Land'), 15 to 20 yards wide, which ran from northwest to southeast through the Troop's sector."

"Enemy 'feeler' patrols, consist of from five to ten men, had been reconnoitering Troop 'C's' area nightly for approximately 5 to 6 days after this unit had relieved the unit previously in the line."

"Up to the time of the special situation, enemy artillery and infantry activity had been considered normal for this particular sector."

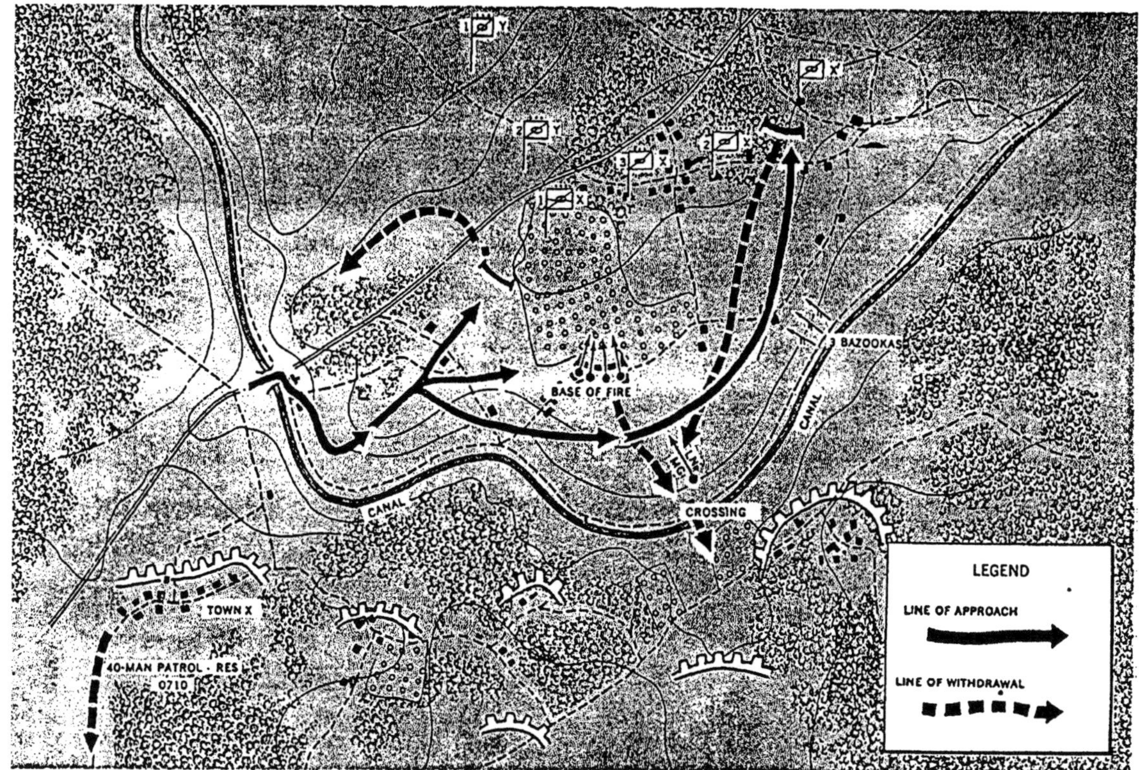

Approach and withdrawal lines of a German patrol attacking U.S. Positions.

SPECIAL SITUATION

"An enemy patrol, later determined to have consisted of 50 to 60 men with a 40-man reserve, crossed the canal at the bridge sometime between 2158 and 2330. Continuous flares and lights were reported in Troop 'C's' area throughout the night (It should be noted that enemy patrols use flashlights to draw fire, and thus disclose the location of dugouts and strong points.) Possible movement of 20mm cannon was reported at 0201 and 0225 from the vicinity of the canal.

"Enemy action started about 0500 with diversionary shelling of the infantry on our left flank. At 0559 the enemy artillery shifted its fire into Captain Stouffer's area, and laid a heavy barrage for approximately 5 to 10 minutes. After the barrage lifted, the enemy patrol moved from its assembly point, and split into segments, as indicated on the accompanying map. Twenty to 25 men, equipped with automatic weapons and five light machine guns, established a base of fire. Twelve to 15 men equipped with automatic weapons and a 20mm cannon on a hand carriage mount constituted the left flank patrol. (Later the 20mm cannon was discovered to be 20mm Flak 38 converted to the ground mount.) Twelve to 15 men, with automatic weapons and bazookas, constituted the right flank patrol Before these elements moved into their positions, they proved extremely proficient in disarming American booby traps, both the prefabricated and improvised types."

"Upon moving into position, the enemy patrols opened small-arms and machine-gun fire on suspected dugout positions. When positive identification of one of our strong points had been established, they shifted to solid tracer fire from their automatic weapons. This tracer fire was targeted on our positions, and then was raised until the tracers were streaming into the sky, making a perfect X. Two or 3 minutes later, enemy artillery would open up on the area where the tracers had crossed. When artillery had registered on one point, the patrols then selected another dug out or strong point, again selected targets by the use of tracers, and again adjusted for the enemy artillery on the south side of the canal."

"Another method of bringing fire upon our strongpoints was to have one man from the enemy patrol place flares, much like our railroad flares, on the spot where he wanted fire, and within a few minutes enemy artillery was registered on that spot."

"The patrol on the right flank was equipped with bazookas. It fired at our dugouts simultaneously with the enemy artillery, and then quickly moved into alternate positions to fire with the next salvo."

"During the action of designating targets with tracer fire, a machine gun, which had been left on the north side of the canal, provided liaison and adjustment for the right-flank and left-flank machine guns. If this machine gun picked out a target, it would fire solid tracer into the area and designate targets, or adjustment for the machine guns of the enemy's right and left flank patrols."

"At 0616 our artillery opened fire on the enemy patrols, firing time bursts. This fire, together with that of the enemy artillery, caused the patrols to become disorganized and to retreat south of the canal. Our artillery swept the canal from left to right for 200 yards for possible crossing points."

"At 0710 and 0730, spasmodic fire from the enemy's automatic weapons covered his retreat across the canal. At 0710 a 40-man patrol was seen moving south from the town of X. It was assumed that this patrol was a reserve for the one which had infiltrated into Troop 'C's' area, and that if the previous patrol had been successful it would have moved across the canal to consolidate its positions and gains. Our artillery opened fire on a fresh reserve patrol, and caused it disperse. As the leader of the patrol fired a single red flare, and the enemy artillery ceased firing."

CONCLUSION

"A *resumé* of the situation revealed that we had only one man wounded, with the others shaken up. A final estimate found that the enemy artillery had dropped approximately 400 rounds into the sector, and that friendly artillery had fired over 300 rounds into the same area."

"The enemy patrols suffered two dead and from 10 to 15 wounded. That afternoon, we sent three special patrols down to the canal, to investigate the area for crossing points. Patrols found two white flags stuck into the ground; the enemy seemingly had used these flags as markers during the night when they returned to the canal."

"Upon inspection of the bodies of the two dead Germans, it was found that their *Soldebuchs* were missing. Either these had been removed from the dead mens' pockets by their comrades, or else it was a policy not to carry these pay books while on patrol. No unit identifications were obtained."

"While German patrols pull every kind of crazy stunt to draw our fire, their night-patrol operations indicate careful planning, excellent cooperation, and precise execution."

"Although this one U.S. Troop has experienced 765 rounds of artillery fire in 11 days, its personnel have sustained four superficial wounds. (This artillery fire has included 10 to 20 rounds of 240mm projectiles.) The unit's careful consideration of experience related by the others in the *Intelligence Bulletin* accounts for its success to date both in repelling raiding parties and in escaping serious casualties."

Intelligence Bulletin Vol III, No 8. (April 1945), pages 72-78.

IN BRIEF

ASSAULT-UNIT (STOSSTRUPP) TACTICS

The principle of active defense has been the key to German tactics on the Western Front. An important aspect of this principle is well illustrated by a memorandum on assault-unit (*Stosstrupp*) tactics, issued early in 1945 by the German High Command.

The purpose of assault-unit operations is to place an Allied force on the defensive, if only temporarily, and during this period subject it to losses of men and material which must be compensated for at the expense of its attacking units. Moreover, the High Command observes, such operations raise the morale of our troops and increase their confidence in their "unquestioned superiority over the enemy."

Using information obtained from reconnaissance, from captured maps, weapons and other material, and -- if possible -- from Allied prisoners, German intelligence informs the commander of an assault unit about the intentions, strength, organization, composition, and fighting quality of the opposition.

Basically, two types of missions are prescribed for assault units: destruction of specially selected Allied positions and the personnel occupying them, and the destruction of Allied units which have penetrated German defensive positions. However, it is pointed out that there always are opportunities for assault-unit operations. In each instance, careful consideration is given to the question of whether the expected success is worth the commitment of trained assault troops. "Weak points in the enemy front always can be found, and these *must* be exploited for assault-unit operations," declares the High Command.

In laying down certain basic guiding principles for assault-unit operations, the High Command orders that qualified men be organized into assault-unit detachments (*Stosstruppenabteilungen*) and be trained in the light of past experience. No standard organization is prescribed, but it is mentioned that each detachment should contain a balanced complement of infantrymen, engineers, scouts, signal personnel, and medical personnel (litter bearers are included, not only to take back the dead and wounded, but to carry captured equipment.) It is pointed out that heavy infantry weapons may cooperate to advantage, as may artillery. The following equipment is suggested: trench knives, sharp spades, hand grenades, pistols with extra magazines, machine pistols (the clips containing 24 rounds only), Very pistols (for dazzling and confusing the opposition) charges to be used in demolishing pillboxes, Tellermines, pole charges, Molotov cocktails, smoke grenades to be used in routing Allied soldiers out of fortifications, and ashes or scraps of paper for marking paths. Raiding parties are made up as many assault-unit detachments as may be necessary. ("Possibly several," the High Command says, vaguely.) However, the Germans emphasize that the mission of such a party must not be too ambitious. "Opportunities that arise in the course of the venture may be exploited to a limited extent by the assault-unit commander,' concedes the High Command. Then follows this very significant observation: "Feints launched at the time of the main breakthrough distract the enemy and cause him to take wrong countermeasures."

Before assault-unit operations, Allied positions are constantly observed and continually reconnoitered.

In deciding upon a mission, the Germans not only take into account the Allied strength and armament in individual sectors, but are strongly influenced by what German intelligence has reported regarding the state of Allied morale in the various sectors. Terrain estimates are studied carefully. Finally, a decision is reached as to whether the assault unit will make a sudden attack, purely infantry in character, based on surprise, speed, and the exploitation of darkness or fog, or whether the unit will try to penetrate Allied positions with the assistance of heavy weapons. The choice between the latter two types determines the tactics and timing of the operation.

An infantry assault based on surprise, and not employing artillery, is regarded as the most practical way to capture prisoners. The High Command remarks that Allied guards frequently are caught sleeping, "because they believe that the Germans avoid night ventures." The first few hours after midnight are recommended as the best for an assault of this type. When Allied forces are alert because of actions on preceding nights, a different time should be chosen for the attack, the Germans add.

On the other hand, the Germans point out, when an assault unit has the mission of penetrating Allied positions, and destroying the positions and their occupants, the cooperation of heavy weapons generally will be needed. An undertaking of this sort usually takes place during the daytime, and the late afternoon hours are regarded as the most favorable. As soon as the objective has been decided upon, detailed reconnaissance is carried out and constant observation is maintained. This work is performed by members of the raiding party themselves, so that they may learn as much as possible about the opposing Allied force, its positions, and the terrain. In all communications the operation is identified by a code name, the day is called "X", and the hour, "Y".

"THE GERMANS COUNTERATTACKED"

U.S. troops who have fought the German enemy have learned that the German defensive doctrine is built chiefly upon the counterattack. This principle of active defense at times has reached the scale of a counteroffensive; however, it is best illustrated in the persistent local counterattacks staged so often by units deployed on a regimental, battalion, or company front.

Two examples of what might be expected from an aggressive defender occurred last year when the U.S. troops were attacking the Siegfried Line near Aachen.

A U.S. division, engaged in breaking through the fortified area, sustained a German counterattack launched by two battalions of infantry, seven tanks, and from 15 to 20 assault guns. The counterattack fell upon the division's left flank, and came shortly after the morning fog had lifted. There was no heavy artillery preparation, and the Germans hit the most vulnerable spot at their most opportune time. The German infantry was stopped by U.S. artillery and infantry action before the counterattack -- a village in the rear of the American line -- was reached. However, some tanks and assault guns did get into the village and caused some trouble before they were eliminated.

This counterattack was one of 17 such assaults the Germans launched during an 11-day period. The units staging these attacks varied in strength from a company to a battalion. Tanks, closely coordinated with the infantry, were used in13 of the actions.

On another occasion, a U.S. division was conducting a successful attack until it was hit hard from two directions by two counterattacking forces, each composed of two German infantry battalions supported by tanks.

The counterattack was conducted by troops from a German division, which had been believed to be far in the enemy rear, and was launched without warning at first light. There were no feints and no artillery preparation, other than what seemed to be a normal shelling of the American positions. However, the American salient in that area was ringed with a horseshoe of German artillery fire, which maintained an effective fire screen both before and during the counterattack.

As a result of these aggressive defensive moves on the part of the enemy, the Allied commanders in that area reached the following conclusions:

With the lengthening nights, and with limited air observation and photography during the day, the division has demonstrated that he can mass a large force -- two divisions with as many as 50 tanks -- in an assembly area close to our lines without any of our forces becoming aware of it.

Then, taking advantage of the morning fog or haze, he can attack and be on us with less than a half-hour's notice. These conditions in proximity of wooded areas greatly increase the necessity for alert observation posts, listening posts, air observation posts, aggressive patrolling and defensive preparations for a variety of eventualities. Rapid, complete dissemination of each bit of information to the next higher echelon frequently can produce the picture of lurking dangers, and disaster may be avoided.

The increase in enemy artillery fire is the most significant change since the Normandy operations. Decreased aerial observation and photography makes it more difficult for our artillery to locate and thus neutralize the enemy's guns. So, greater emphasis on accurate shell reports, and pointed interrogation of prisoners of war as an aid to location of batteries, is needed.

The German's selection of the swamps west of the Meuse as a spot to employ two of their best mobile divisions is a startling remainder that the enemy cannot be trusted always to attack according to the "book." They remain clever, aggressive foes.

MODEL'S METHODS

After the first day of the Ardennes counteroffensive, Field Marshal Model issued an order analyzing the U.S. defenses and setting forth certain attack principles to be followed by the German forces.

Quick exploitation of the successes of the first day of attack was urged. Model regarded such exploitation as decisive. The first objective, he said, was to achieve freedom of movement for the mobile units.

To this end, the U.S. main zone of resistance was to be broken through at its weakest points. These points were to be discovered by reconnaissance in force along the entire front.

The weakest strongpoints were to be taken by encirclement, with simultaneous concentration of fire by artillery and heavy infantry weapons, and by the use of tanks or assault guns. The neighboring strongpoints were to be neutralized by fire or by smokescreens.

Whenever possible, strongpoints were to be bypassed, and left for the reserves to deal with. "Exaggerated attention should not be paid to temporary threats to our flanks," Model said.

The main effort in antitank defense was to be undertaken by the leading units. The units were to be in a position to defeat U.S. reserve units quickly. The remaining antitank weapons were to be distributed in depth throughout the attacking or advancing units.

"Artillery preparation for the attack must not be handled in a shortsighted manner," Model declared. "If, for some reason, the infantry is delayed and cannot break in after the last round of a period fire has been delivered, observed fire must be continued until the infantry is ready to break in."

Intelligence Bulletin Vol II, No 7. (March 1944), pages 61-63.

Section III. A GERMAN COMPANY IN THE DEFENSE

1. INTRODUCTION

Recently a Panzer Grenadier company commander, troubled by the loss of life and matériel that his unit had been suffering in Italy, made an effort to alter certain methods that his men had been employing in the defense. His attempt is significant, in that it is an instance of a German commander undertaking hastily to revise his company's practice, in the light of experience freshly acquired in fighting United Nations forces. Much has been said about the rigidity of German junior leadership in the field, and not enough, perhaps, about its adaptability. The illustration at hand shows how a German junior officer tried eleventh-hour measures in the hope that his unit might avoid further reverses.

2. THE COMPANY COMMANDER'S INSTRUCTIONS

a. It is stated that, since a creeping barrage always preceded an attack, this type of fire was to be a signal for each man to go at once to his alert station and make a further brief check of his weapons.

b. Even during the barrage, every man was to keep a continual lookout, frequently raising his head above the parapet. This was described as particularly important when the fire moved, or "lifted", because, it was said, a hostile advance would follow the barrage closely, and the opposition would use mortar fire and grenades for purposes of deception.

c. When the attackers arrived in close proximity to the position, special attention was to be paid to any cover or dead space within hand-grenade range. Hand grenades were to be used against any hostile soldiers who might succeed in reaching such places.

d. The first section to discover that a hostile attack was in progress was to send a reliable, speedy runner to platoon headquarters by the safest route. It was stressed that speed was essential if the heavy weapons were to give proper support.

e. It was ordered that the position be held at all costs. Every man was to stay at his post and fight. A single well-aimed shot was to be regarded as more worthwhile than a badly placed burst of machine gun fire.

f. The dispatch of a runner was not to be considered when a Very pistol was available. The following signals were to be employed:

Red .... Attack by hostile force.
Green .... Artillery barrage has lifted.
White .... We are here.
Violet .... Tank Attack.

[It should be remembered, of course, that all German signals may be changed frequently.]

g. It was stated that platoon headquarters should have sufficiently good observation to enable the platoon commanders to keep up with the situation and to insure against any hostile attack achieving surprise.

h. Platoon headquarters were to be turned into strong points, so that a hostile force could be engaged at any time from the depths of the position. A reminder as to the effectiveness of enfilade fire was added.

i. It was ordered that, if the next hostile attack were to be made at night and with armor, the forward sections were to fire with everything they had, while the best hand-grenade throwers were to be assigned for this specific duty. It was pointed out that the resulting damage to the opposition's morale might serve to halt the advance.

Intelligence Bulletin Vol III, No 9. (May 1945), pages 52-58.

Discovered in Combat

"SPIDER HOLES"

"'Spider holes' are what we call ground positions that German snipers sometimes use. These holes are 2 1/2 to 3 feet in diameter and approximately 6 feet deep. Such a hole is covered with a camouflaged lid made of criss-crossed sticks tied securely together with wire or cord, with a top layer of sod. The sod is neatly cut and trimmed to match the surrounding terrain. Thin wire holds the sod in place over the criss-crossed sticks. A sniper can tilt the lid slightly in order to observe his surroundings; if he chooses to, he can raise the lid and slide the whole thing aside."

"To men who have been accustomed to watching trees and other elevated places for signs of snipers, these ground positions may be a totally new thing, and a G.I. may be caught off guard."

"The sniper's pet tactic is to fire on the first column coming up, and do all the damage he can. When the U.S. fire gets too hot for him, he closes the lid and waits for the first column to pass. Then he surrenders to the next column. It is difficult to detect the sniper's 'spider hole' because of its excellent camouflage."

USE OF RICOCHET FIRE

"In Italy the Germans often have employed ricochet fire in terrain honeycombed with stone walls. They use this tactic to pin us down and delay an advance. I'll give you an example of ricochet fire used in this manner."

"My platoon was moving down a narrow Italian road, and had approached to within 25 feet of a stone wall that ran along the far shoulder of a left-hand curve in the road. At this point the road made about a 75-degree bend to our left. It was morning, and visibility was good. The road beyond the bend had been scouted, and no evidence of enemy had been observed. Small trees grew in clusters at irregular intervals along the road. Flat farmland bordered each side of the road to a path of about 50 yards, and then rolling hills gradually ascended to a height of approximately 75 feet."

"The leading platoon was marching about 25 yards ahead of the second platoon, as a security measure. As my platoon began to enter the curve, a heavy stream of machine-gun bullets spattered against the wall on our right. Within a few seconds, ten of my men lay dead or wounded in the road. A strong patrol was dispatched immediately to knock out the enemy machine-gun nest that had subjected us to this ricochet fire. When the patrol returned after destroying the machine-gun nest, it was reported that the German machine gunners had been operating from a small hill overlooking the bend in the road.

At a range of about 100 yards, the Germans had zeroed their guns on the middle of the stone wall."

"The machine-gun bullets become dum-dummed when it strikes a hard, smooth surface, and the slug has a tearing effect on personnel. Because of the inherent inaccuracy of this ricochet type of firing, it is difficult to judge the extent of the area affected by the rebounding bullets. The pattern is highly irregular."

MINING SUPPLY PATHS

"The Germans in Italy have been very skillful at crawling through our lines at night, usually moving via our supply ditches. They have a habit of laying mines in our supply trails and then crawling back to their own lines. Sometimes they fail to get back; on several occasions we have discovered these night crawlers as trying to hide themselves under rations and water."

MORTAR POSITIONS

"From a German operations map, we learned that we had been looking for the enemy's mortars too close up. His 120mm mortars were to the rear of his artillery positions."

MACHINE GUN TACTIC

"Our company was moving toward Arnhem, after having captured our objective at Veghel, in Holland. At 0200 my squad and another squad were ordered to take a German position consisting of what was thought to be a machine gun and 11 riflemen, all dug-in around an old house near an orchard."

"When my squad had advanced within 25 yards of this position, the enemy opened fire with three machine guns instead of one. Two machine guns, one on each flank, fired tracers at approximately head-high elevation. The gun in the middle delivered grazing fire about 6 inches above the ground, and did not use tracers. As a result the center machine gun was extremely difficult to locate, and we were unable to take the position."

HOW THE GERMANS RECAPTURED A TOWN

"Our company moved into the town of X, in France, with little opposition from the Germans. The enemy had four antitank guns protecting the road leading into the town, but these guns were easily knocked out by rifle fire when the Germans made the mistake of allowing our small-arms fire to get within range of the antitank guns. After our company had moved into the town, our men were placed in houses chosen for this purpose. There was no systematic rounding-up of the civilians."

"At 2100 on the same day, the Germans counterattacked. Their tactics consisted of sending approximately one company of infantry into the outskirts of the town, apparently as an advance element. Two Panther tanks stayed outside the town, beyond the range of rifle fire, and fired their 75mm shells into the area that our troops were occupying. These tanks fired with remarkable accuracy, and aimed only at those buildings in which our troops had been placed. It is my belief that the civilians in the town had informed the enemy as to the exact whereabouts of our troops."

"Since the tanks stayed out of range of our small-arms fire, there was little our company could do to put them out of commission. Small patrols were sent out, but they were unable to get through the German infantry, and thus were unable to contact the tanks. Nor were we able to get our bazookas within range of the armor."

"The Germans kept up fire from the 75's, and eventually forced our company to retire from the town."

Intelligence Bulletin Vol I, No 8. (April 1943), pages 1-14.

Section I. GERMAN COMBAT IN WOODS

1. INTRODUCTION

The following German document summarizes lessons learned by the German army in Russia, with regard to combat in wooded terrain. It is of special interest to American troops, inasmuch as the Germans may reasonably be expected to utilize these lessons in other theaters in which large woods, often swampy and thick with underbrush, are to be found.

2. THE DOCTRINE

a. Principles of Leadership

(1)   Wooded terrain often enables us [German forces] to advance within assault distance of the enemy, to bring up reserves and to shift forces to the critical point of attack (*Schwerpunkt*). Wooded terrain also is favorable for coming to close grips with enemy tanks.

In woods, it is practicable to seize the initiative against an enemy who is superior in heavy weapons, artillery, and tanks. By using surprise attacks, it is possible to annihilate such an enemy, or at least to maintain a successful defense against him.

(2)   Difficulty of movement through wooded terrain, and of observation over it, demands that heavy weapons and artillery be attached to the combat units they are to support.

(3) In woods fighting, observation difficulties and a general lack of knowledge of the situation call for courage, tenacity, and the ability to make quick decisions. Mobility,

coupled with alert, cunning leadership on the part of all commanders, can play a decisive role.

(4) In this type of combat, everything depends on the concentrated energetic employment of the infantry strength with a view to annihilating the enemy. Systematic fire preparations and the development of protecting barrages are seldom possible. Also, there will usually be many blind spots in defensive fires.

Therefore, the number of rifles and machine guns is often the decisive factor in woods fighting.

(5) Surprise is even more important in woods fighting than in combat in open terrain. For this reason, woods fighting demands, above everything else, careful planning and silence during all movements.

(6) In woods fighting, the necessity for sending out strong patrols to the front, flanks, and rear can lead to a dangerous scattering of strength. In situations where there is danger of being cut off or surrounded, it is generally preferable for the commander to hold his strength together. Such situations arise frequently, especially with small units. This must not lead to hurried or ill-advised measures, or to panic. Caution, determination, and skillful employment of available forces will generally permit offensive actions in which the enemy can be defeated or annihilated.

(7) Movement and combat in woods demand formations in depth. This facilities control of forces, mobility of leadership, rapid transition of orders, and readiness to deliver fire quickly when a flank is in danger.

(8) An advance based on gaining one intermediate objective after another, and on reorganizing the units after each objective has been reached, protects against surprise and makes control easier.

(9) To coordinate the effort in woods fighting, the commander must formulate a detailed plan of operation, and must give each subordinate unit a definite mission. Often he must describe in detail how the mission of a subordinate unit is to be accomplished.

(10) Combat in extensive woods, especially in enveloping movements and in encirclements, often consists of a series of small fights. The individual assault groups must act as a coordinated whole, despite difficulties in the transmission of orders and in communications between units. Commanders of participating units must understand both the mission and the situation.

(11) Any commander who is forced by the situation or the terrain to depart from the prescribed plan of operation must obtain beforehand the approval of his higher commander. This is necessary so that the latter can coordinate the proposed action with that of other units in the woods, and above all with the fire of the heavy infantry weapons, the artillery, and the German Air Force. Such coordination is necessary in order to avoid losses through the operation of friendly supporting fire.

(12) Since the German Air Force is often unable to obtain adequate information about the enemy, and since it is seldom practicable to employ the motorized and armored reconnaissance units on a large scale, the employment of numerous strong scout patrols on foot becomes highly important.

b. Reconnaissance, Observation, Orientation

(1) In operations in wooded terrain, all our [German] troops units must carry out ground reconnaissance continuously to avoid surprise attacks by the enemy.

In general, several patrols operating abreast of each other should be sent out to the front. Other patrols should operate along the flanks. In this connection, it is important to make the distance between adjacent patrols wide enough to avoid the danger of one patrol being confused by the noises made by a neighboring patrol In woods with thick underbrush, this distance should be at least 160 yards.

(2) In line with the principle of silent movement, the equipment of patrols must be tested carefully, and anything which creates a noise or is too unwieldy must be left behind.   The armament of patrols consists of machine pistols, rifles (if possible, automatic rifles equipped with telescopic sights), and many *Eier* (egg) grenades. Since the ear must be constantly alert, the steel helmet may be left behind.

(3) Patrols should obtain information which will answer such questions as the following: Where is the enemy, or where is he believed to be? Where are his left and right flanks? Where are his advance security elements? What are the habits of his patrols? Where are his fields of fire?

If contact with the enemy is effected, it is desirable to obtain early information regarding gaps or weak points in his positions. This provides a basis for quick tactical decisions by the commander.

It is especially necessary for reconnaissance missions to obtain information regarding existing roads, paths, and clearings; ditches, streams, and bridges; and such characteristics of woods as the thickness of underbrush, the height of trees, the location of high or low ground, and the location of swamps.

(4) The commander of companies, platoons, sections, and squads must always detail special lookouts to protect against snipers in trees. Individuals who spot tree snipers are to dispose of them by aimed rifle fire. Sweeping the treetops with machine guns will be resorted to only in cases where enemy snipers cannot be located definitely.

(5) During halts, observation from treetops is profitable.

(6) Patrols in the woods must carefully observe paths and trails.   Important conclusions concerning the position of the enemy can be drawn from the location and condition of these trails. It is important to note whether or not a trail has been used recently.   One way of judging this is to inspect the morning dew for any unnatural disturbance.

(7) If there is no terrain feature on which the patrol can orient itself, the compass must be used. Each patrol must carry at least two compasses, one for the leader and one for his second-in-command. The leader is in front; the second-in-command brings up the rear, guarding against any deviation from the proper course.

c. On the March

(1) Our [German] troops must be prepared to erect short, strong bridges and to lay corduroy roads. Engineers must be placed well forward to clear the way and to remove obstacles. Also, many road-working details must be provided to assure mobility for all units.

(2)   When paths and roads through swampy woods must be used, it is especially desirable to locate local inhabitants as guides. Routes of this type often are not shown on maps.

(3)   In woods fighting, the increased length of time required for bringing forward various elements from the rear demands that the advance guard be made very strong. Heavy weapons and artillery, staffs, and signal detachments are to be placed well toward the front.

(4)   All elements of the column must be prepared for simultaneous defensive action, and must expect surprise attacks against flanks and rear.

(5)   As a rule, flank guards and rear guards must be used. The flank guards must be able to leave the roads and move across country. The strength of the flank guards, as well as the interval between them and the main body, depends on the strength of the main body, the character of the woods, and above all on the character of the road nets. Lest flank guards be cut off and destroyed, they must guard against operating too far from the main body.

(6)   It is necessary to include, in all echelons of the column, detachments which have the mission of searching for enemy tanks.

(7)   Strong air attacks, artillery fire, attacks by guerrillas, or attacks by enemy troop units may be make it necessary to conduct the march off the roads while passing through woods.

(8)   In clearing road blocks rapidly, it is advisable to attack the blocks frontally with fire delivered from units on each side of the road. This fire will pin the defenders to the ground. Meanwhile other elements of the attacking force should envelop the roadblock from the rear.

d. Approaching the Enemy

(1)   When our [German] reconnaissance has indicated that enemy resistance is to be expected on the route of march, and when contact with the enemy seems near at hand, it is often wise to abandon the march on the road fairly early, in order to gain surprise and launch an attack from a direction that is tactically favorable.

(2) If the woods are thin, the advance formation can be loose, with wide intervals. If the woods are thick and relatively impassable, the troops must be held close together, echeloned in depth.

(3)   Units should move by bounds when advancing through woods. Orders issued in ample time should specify the successive objectives. These must be clearly defined features, such as roads, paths, creeks, and so on. After reaching an objective, the unit must halt long enough to reorganize and re-orient itself, to let the heavy weapons and artillery weapons catch up, and to take new security measures.

(4)   It has proved a good idea to provide special close-in security elements between the main body and the normal security elements. The special elements consist of small groups of infantry, equipped with close-combat weapons -- especially machine pistols.

The main body advances in deep formation, with security to the flanks and rear, as described above.

Mortars, antitank guns, and heavy infantry weapons should be placed well forward, immediately behind the leading infantry. This is done so that unexpected enemy resistance can be broken at once by heavy fire.

(5) Short halts should be called on reaching clearings, roads, and paths, and on leaving the cover of woods. Machine guns and heavy infantry weapons are brought up to cover the advance across open land. Patrols move through the woods to the right and left, so as to reconnoiter the edge of the woods on the opposite side of the clearing. When the advance is resumed, the open clearing is to be avoided even if the edge of the opposite woods is reported to be free of the enemy. Clearings which cannot be avoided are to be crossed in swift bounds.

(6) When German troops are within sight and range of the enemy, further advance is made by creeping and crawling so as to come within close combat distance. Even under strong enemy fire, creeping and crawling through woods can be accomplished successfully.

e. Attack (General)

(1) To achieve surprise, we [German forces] should employ all available weapons so that the enemy will be deceived as to our plan and strength and as to the time and place of our attack. Feint attacks may be made in the woods and with weak forces. Noise alone may serve the purpose. These measures confuse the enemy, tempt him to employ his reserves prematurely, and therefore weaken his power of resistance. If possible, the attack should be made so that the enemy is enveloped from both flanks, or at least struck in one flank. (An enveloping attack by the enemy can best be repulsed through the employment of forces brought up from the rear.)

(2) In thick woods, natural features which run perpendicular to the direction of the attack serve well as objectives. The stronger the expected enemy resistance, the shorter must be the distances between objectives.

(3) Surprise will be obtained chiefly by the manner in which fire is opened. Generally, the commander himself will give the order to fire. Fire must be coordinated and delivered in short, heavy bursts; this has a useful psychological effect in the woods.

(4) The fire of the defender, which normally strikes the attacker at very short ranges, must be traversed rapidly, regardless of consequences. Experience has shown that this method results in fewer losses than if the attacker goes into position and overcomes the defender by firepower.

(5) In general, it is pointless to deliver fire on the enemy after he has abandoned his position. Through rapid, determined pursuit, he must be prevented from reorganizing and gaining time to launch a counterattack. However, if fighting has been heavy and has involved much man-to-man action, it is profitable to halt briefly after the enemy collapses so that we can reorganize and concentrate our strength.

(6) Since trees and underbrush reduce the effect of individual rounds, woods fighting calls for the use of more ammunition than does combat in open terrain. Therefore, the problem of ammunition supply requires special consideration.

(7) In woods it is usually impossible to carry on an attack after nightfall. For this reason, troops must break off the fighting in ample time to reorganize a defensive position for the night.

f. Attack against a Weak Enemy

(1) We [German forces] are most likely to succeed in an attack against a weak enemy if our approach is noiseless, and if the assault is launched from the closest possible distance on either enemy flank.

(2) When our patrols have developed the possibilities for an envelopment, troops are brought up from the rear. While they envelop the enemy flanks and rear, our forward elements attack energetically to the front.

Frequently the patrols sent out to locate the enemy flanks, or the messengers sent back by these patrols, can serve as guides for our enveloping forces.

The units assigned to the enveloping force will be controlled by their leaders through prearranged sound signals.

The attacking elements in front of the enemy positions open fire and launch their attack, with battle cries and with bugles blaring "the charge."

(3) In general, the commander belongs with the elements attacking to the front, since this enables him to judge the situation and make decisions regarding the employment of rear elements.

g. Attack against a Strong Enemy

(1) This type of attack follows our [German] principles of attack against organized positions. Assault detachments are formed, equipped with such weapons of close combat as hand grenades, smoke grenades, and Molotov cocktails. Flame throwers are especially effective in the woods.

(2) Assault detachments seek out the weak points in the enemy position, and try first to effect a small breach. If the woods are thick, and if gaps in the enemy positions are discovered, it is advisable to infiltrate silently into the position with small detachments. These detachments attack enemy centers of resistance and security posts, annihilate them, and attempt to throw the enemy into confusion. They prepare the way for the attack by the main body.

(3) Woods often prevent the opposition from detecting our advance. This makes possible the assembly of our attacking force at points closest to the enemy position, especially just before dawn.

(4) A sudden and determined surprise attack without preparation by fire is usually more effective than an attack preceded by fire preparation.

(5) The fields of fire prepared by the enemy must be avoided. Machine guns, antitank guns, and individual artillery pieces must be brought into position in these fields of fire and must silence enemy guns.

(6) The spearhead of our attack must penetrate deep into the enemy position, despite all difficulties and lack of observation. The rear elements of the attacking force widen the penetration and mop up the position.

h. Support by Heavy Infantry Weapons and Artillery

(1) Since the ranges are usually very short, our [German] heavy machine guns will often be employed in the role of light machine guns. Light machine guns may be brought into action rapidly, and may be shifted from position to position very readily. Ammunition supply vehicles are brought forward by bounds.

Heavy infantry mortars are attached to the heavy infantry platoons. Use of smoke shells during siting has often proved profitable. Care must be taken to insure that the trajectory of the mortar shell will not be blocked by trees or foliage.

The mobility of light infantry weapons and light antitank guns makes them extremely useful; they will normally be attached to the infantry companies and will be employed in direct firing.

The use of armor-piercing shells by antitank guns is effective against targets of all kinds, since these shells are not easily deflected by trees or foliage. On the other hand, our "hollow-head" shell (*Hohlkopfgeschosse*) is less successful, because it is sensitive and explodes prematurely when it strikes a tree or bush.

(2) Because of observation difficulties in the woods, our artillery is likely to have great trouble with shells falling short of enemy positions.

Many advanced observers must be attached to the leading companies. Whenever observation is possible, fire must be opened suddenly.

The laying of telephone wire takes time. Therefore, reconnaissance units with the leading elements must be equipped with other means of communication, preferably radio.

In favorable terrain it is often possible to make use of observation from high points outside the wood. Flares are used as signals between the observer and the infantry commander to indicate the location of the leading elements, to define the targets, and to control fire.

To match the advance of the infantry, artillery fire is moved from object to object. Certain phase lines will be prescribed, and fire will be directed on these at the infantry's request. Short violent salvos are especially effective.

i. Mopping Up a Woods

(1) When we [German forces] attempt a rapid mopping up of a woods, we normally succeed only through the employment of forces moving in different directions to surround the enemy.

(2) The mopping up of woods by the "combing" method -- in which our troops advance in a line along a broad front, with only a few yards between individual soldiers -- has proven ineffective. There is always a danger that the enemy will concentrate his forces at a given point and break through our own line. Therefore, we must keep our forces concentrated and, according to the situation, direct strong and compact assault detachments along the roads and paths -- with all movement following a carefully prepared plan.

(3) To forestall enemy attempts to break out of the woods, we must cover the edges of the woods with infantry weapons and artillery, and must employ tanks and self-propelled artillery.

(4) Against an encircled enemy, harassing fire in ever-increasing density and the employment of air power are especially effective.

Advance observers, attached to the assault detachments and equipped with individual radio sets, can direct the artillery fire and the air bombing so that there will be no danger to our own troops.

j. Defense

(1) When defending in a woods, the danger from surprise attacks is increased We [German forces] must not wait until the enemy, taking advantage of the abundant cover, launches an attack from the immediate vicinity. We must seek out the enemy, attack him, and annihilate him.

(2) Mobility in the defense is the best means for deceiving the enemy as to our own strength and intentions. This mobility often leads to the defeat of a superior enemy force.

(3) It is especially important to build up centers of fire quickly and to employ reserve strength, even in small units, with a view to annihilating the enemy through counterattacks. Heavy infantry weapons, artillery, and reserves must be held close to the point of likely action.

(4) Organization in depth and complete cover of the front by fire are seldom possible, even when strong forces are available.

Woods enable a defender to erect many effective obstacles in great depth. These obstacles will often stop the enemy or canalize his attack in a direction favorable to the defender.

(5) The employment of the tank-searching details close to the enemy's best routes for tank approach is highly profitable.

(6) When a complete defensive position cannot be organized due to lack of time or manpower, a continuous strong obstacle must be erected -- and, if possible, many centers of resistance, capable of all-around defense and of holding under fire the enemy's avenues of approach.

(7) The edges of woods normally lie under enemy fire and therefore are not to be occupied. Our weapons must operate from within the woods. If observation permits, they should be at least 30 to 50 yards from the edge.

Each center of resistance must be protected all around by minefields. Within the center of resistance, an adequate supply of hand grenades must always be within reach.

Every effort must be made to clear fields of fire. These must be located so that our fire will strike the attacking enemy's flank.

(8) Camouflage of the position against observation by treetop observers is essential, but uniformity in the organization of positions and in the methods of camouflage must be avoided.

Camouflage screens are to be used at all times. Branches and twigs used for camouflage must be renewed every morning (Dry or withered foliage will betray even the best located- positions.)

(9) Paths and trails between individual centers of resistance, and to the rear, are to be provided and marked. These paths are necessary for mutual support by adjacent units and for quick employment of reserves.

Paths must be kept free of dry wood and foliage so that the patrols will not be betrayed by crackling and rustling noises. The tendency of soldiers to take short cuts, and therefore make new paths, must not be tolerated.

Patrols must never operate on regular schedules or by identical paths. The enemy will soon discover the routine and will annihilate careless troops.

(10) Defenses in the woods demands the employment of numerous observation posts for the artillery (3 to 4 per battery). For this purpose, signal equipment from the division signal company must be made available. Many heavy barrages must be provided in front of the main line of resistance, and especially between the gaps in the centers of resistance.

The gun positions of the artillery must be protected against close-in attack. To this end, the position must be surrounded with centers of resistance, especially to the flanks and rear, and with increased numbers of security patrols.

These provisions are especially important in cases where only weak infantry forces are available and where, therefore, the development of the defensive position in great depth is not possible.

(11) Obstacles must be erected in front of every position, especially along rivers and creeks. The enemy may use the latter as avenues of approach. All obstacles must be covered by fire, even if only by patrols. Wire and booby traps are to be used liberally. Alarm devices, which can be made from captured wire and cans partially filled with stones, should be installed in the wire and obstacles in front of a position. The response to alarms should be rapid. Indeed, each commander must realize that speed can be decisive in a counterattack.

(12) Listening posts and standing sentries are to be employed at all favorable approaches to a position. The hour for relief and the place of relief are to be changed frequently.

(13) The laying of telephone wires, even to the smallest advance unit and neighboring units, is important. (Captured matériel will often be useful in this connection.) Wires should be strung through the treetops.

Intelligence Bulletin Vol III, No 5. (January 1944), pages 65-71.

HOW THE GERMANS FIGHT IN
WOODED AND BROKEN TERRAIN

The Germans recognize that operations in wooded and broken terrain require special combat methods both in the attack an in defense. In such terrain the Germans try to control all roads and trails, so as to ensure the movement of support weapons and supplies. The heaviest fighting therefore generally takes place in the vicinity of these roads and trails.

"In the defense it is considered essential to block roads and trails. Snipers are posted in trees. Centers of resistance are established at curves and defiles, and whenever a road climbs to higher ground."

GENERAL PRINCIPALS

In the attack the Germans maintain careful protective fire as they advance along the roads and trails; when they are obliged to move across open stretches, this protective fire becomes continuous. Roads are opened up as rapidly as possible, and are covered with antitank guns. Special attention is paid to the formations adopted during movement and in battle, to correct employment of fire power, to appropriate communication methods, to the problem of maintaining direction, and to supplying forward elements with an adequate amount of ammunition.

In the defense it is considered essential to block roads and trails. Snipers are posted in trees. Centers of resistance are established at curves, bends, and defiles, and whenever a road climbs to higher ground. Firing positions are prepared just off roads and trails, to command open fields of fire.

METHOD OF ADVANCE

In the approach march, squads and platoons advance on a narrow front, deployed in depth along roadside hedges and scrub growth, and in hollows running in the desired direction. The leading squads, on contact, serve as scouts and patrols. They advance in extended order, with a light machine gun leading. While the squads immediately behind the forward position deploy less deeply at intervals of 30 to 40 paces, the subsequent squads follow in squad columns so as to have all-around observation and protection. Special observers are detailed to watch out for tree snipers.

The Germans believe that when battle is joined, the same formations employed during the approach march should be maintained as far as possible. Fire cover is provided by the support weapons, especially the mortars, which

advance with the forward deployment of squads and platoons as may be necessary. It is a German principal that after resistance has been crushed and hostile strong points eliminated, the original formations should be resumed immediately.

The reserve platoon advances, employing the same close formation, in the rear of the platoon, which gains the most ground. The commander of the reserve platoon arranges for all-around protection, particularly to repel surprise attacks, which may be made by hostile forces from centers of resistance not yet engaged. These protective measures also include protection of the rear.

In the approach march, squads and platoons advance on a narrow front, deployed in depth along roadside hedges and scrub growth.

USE OF FIRE POWER

To eliminate centers of resistance, the Germans employ all available light and heavy weapons, especially mortars.

Since observation in close country is difficult, the Germans not only keep their support weapons well forward, but often use their heavy machine guns as light machine guns.

Terrain conditions are likely to have a definite effect on German employment of mortars. Sometimes observers can work only from treetops. Every effort is made to place observers close to the mortar positions so that corrections can be passed accurately and rapidly to the mortar detachment.

"Sometimes observers can work only from treetops."

"In the heat of battle, disk signaling is preferred."

The employment of message runners is not considered practicable in the heat of battle; instead, disk signaling is preferred. The Germans try not to site their mortars too close to the roadside scrub growth.

The commanders of support weapons are required to report their availability to the leading rifle company commanders and his platoon commanders, and to remain in their vicinity.

The antitank guns follow without orders in the rear of the infantry, as soon as the roads have been cleared. Their principal mission is to take over the job of preventing hostile tanks from using the roads. In addition, so far as their principal mission permits, the antitank guns take part in attacks on Allied centers of resistance, using antitank high-explosive shells.

"The antitank guns take over the job of preventing hostile tanks from using the roads."

Protected by the fire of the support weapons, the infantry works its way forward as close as possible to the Allied centers of resistance. As soon as the support weapons cease firing, the infantry breaks through, hurling hand grenades. The Germans are scrupulously careful in regulating the time when the support weapons are to cease firing -- first the medium mortars and then the heavy machine guns -- and the time when the breakthrough is to be attempted. The points at which the breakthrough is to be made are sealed off on the flanks by squads especially detailed for this job. Hostile positions along hedges or other roadside growth are mopped-up after the breakthrough.

MISCELLANEOUS PRECAUTIONS

Platoons and squads detail men for the express purpose of maintaining contact with neighboring units. These men indicate the headquarters of their own units by means of pennants and by signaling with lamps to flanking squads and platoons. It is a rule that pennants marked "Front Line" must always be put up. Identification panels are laid out, when necessary, to indicate the advance of the front line.

Because the opportunities for unobserved movement are very good in terrain of this type, the Germans make considerable use of runners. Radio-telegraphy and smoke cartridges also are used, in addition to light signals.

"Compass directions are issued before the departure."

Higher headquarters are continually kept informed about the situation, to permit smooth coordination of the attack.

Since the problem of maintaining direction is difficult in closely wooded and unevenly wooded terrain, squad leaders are given specific rendezvous on roads and paths. Compass directions are used before the departure.

Because of terrain difficulties, the Germans find it useful to equip squads with ladders, axes, good knives, and sharp spades. Since ammunition supply is likely to be slow and cannot be relied upon, a generous quantity of ammunition, including hand grenades, is issued to them before their departure.

Intelligence Bulletin Vol I, No 11. (July 1943), pages 18-26.

Section III. COMBAT IN HIGH MOUNTAINS, SNOW, AND EXTREME COLD

Official German military doctrine dealing with the tactics to be used in high mountains and under conditions of extreme cold is summarized in this section. The information has been extracted from German Army documents.

1. IN HIGH MOUNTAINS

a. Command

The German army emphasizes that the skill and leadership of junior commanders are severely tested in mountain warfare inasmuch as forces will generally be split into relatively small groups. The efficient handling of these groups demands a high standard of training and discipline. Columns will often be separated by wide areas of difficult country, and, since lateral communication is often very difficult, command of deployed units becomes much more complicated than during operations over ordinary terrain.

b. Movement

(1) The Germans recognize that the limited number of trails and the tendency of men and animals to become exhausted have a decisive influence on movement.

(2) Units are divided into numerous marching groups, none of which is larger than a reinforced company, a gun battery, or an engineer platoon. The Germans maintain that in this way the danger of ambush can be overcome, and each group can fight independently.

(3) Engineers are well forward (with protective patrols) to help repair roads.

(4) Aware that small enemy forces can hold up the advance of a whole column, the Germans consider it necessary to have single guns well forward. They also regard flank protection as very important; for this reason stationary, as well as mobile, patrols are used.

(5) When unusually steep stretches are encountered, infantry troops [probably reserves] move forward and disperse themselves among the pack animals of the artillery, for the purpose of helping the artillery in an emergency.

(6) Pack artillery moves at march pace (2 1/2 miles per hour) and after marches of over 6 hours, 3 to 4 hours rest is necessary. Short halts are considered useless, because men and animals must be able to unload.

(7) The Germans stipulate that for every 325 yards ascent or 550 yards descent, 1 hour should be added to the time which would be estimated necessary to cover that same map distance on ordinary terrain.

c. Supplies

(1) The supply echelon may include some, or all, of the following means of transport: motor vehicles (with preference given to vehicles of from 1 to 2 tons), horse- and mule-drawn vehicles, railways, pack animals, and manpower. Transport aviation may also be used if terrain permits.

(2) The Germans have found that a cart drawn by two small horses can be highly practical in mountainous country.

(3) The German army stresses that supplies must be initially packed in containers suitable for pack transport, in order to avoid a waste of time in repacking en route.

(4) Man loads vary from 45 to 75 pounds.

(5) In general, supplies are organized into valley columns and mountain columns. Valley columns carry supplies for 2 days and mountain columns carry supplies for 1 to 2 days.

d. Weapons

(1) Light machine guns are used more often than heavy machine guns.

(2) Mortars are used extensively, and often replace light artillery.

(3) Antitank guns and heavy machine guns are mostly used for covering road blocks.

(4) It is a German principle that effectiveness of artillery fire from valleys depends on the careful selection of observation posts, and on efficient communication between these posts and single gun positions. "It cannot be overemphasized," the Germans say, "how difficult it is for artillery to leave the roads or level ground."

(5) In general, the emphasis is on the lighter weapons.

e. Reconnaissance

Apart from normal reconnaissance tasks, it is considered important to mark trails to show: which areas can be observed by the opposition, how far pack transport can be used, where trails need improving, and where troops must assume the responsibility of carrying everything themselves.

f. Signals

(1) The Germans use radio as the primary means of communication, because of the great difficulty of laying lines.
(2) Motorcycles and bicycles are used in the valleys.
(3) The Germans take into account the fact that lateral communication is often very difficult and sometimes impossible.

g. Engineers

German engineers in mountain units are assigned the following tasks in addition to their normal duties: bridging swift mountain streams, clearing mule trails, and constructing rope and cable railways.

h. Attack Tactics

(1) Attacks across mountains usually have subsidiary missions of protecting the flanks of the main attack (usually made through a valley), working around the rear of the opposition, or providing flanking fire for the main attack.
(2) It is a German axiom that the early possession of commanding heights is essential to the success of forces moving along the valleys.
(3) Generally, the German main attack follows the line of valleys, which alone gives a certain freedom of movement to a strong force and the necessary supply echelons.
(4) German troops attacking uphill are always on guard against falling rocks and possible landslides caused by supporting artillery fire.
(5) The Germans have found that attacking downhill, while easier for the forward troops, often presents tactical and ballistic problems for the artillery.

i. Defense Tactics

(1) German officers are reminded that defense of any large area of mountainous country ties down a very considerable number of troops.
(2) The Germans believe that if a crest is to be defended, it is better to have only the outpost position on the crest or forward slope and to have the main line of resistance, with heavy weapons, on the reverse slope.

j. Training

(1) In general, the basic training of German mountain troops is that of normal infantry units. Specialized training comes later.
(2) Battalion officers are trained mountain guides, and must pass tests annually.
(3) All guides are required to be expert at map reading and use of altimeters, at judging weather conditions, at recognizing dangers particular to mountainous country, and overcoming great terrain difficulties in order to reach observation posts.
(4) The necessity for noiseless movement is emphasized, inasmuch as under certain conditions sound may travel farther in mountainous terrain than in open country.
(5) It is stressed that ammunition must be used economically.
(6) Since troops are likely to be separated from their units for a number of days, the Germans require a high standard of discipline and physical toughness.

2. IN SNOW AND EXTREME COLD

a. Movement

(1) *Marching.* - The Germans consider it important that clothing should not be too warm. Weapons are covered. Advance guards are strong, and heavy weapons and artillery are well forward. Antitank weapons are distributed along the column. Ski and sleigh troops may be sent out to guard the flanks. Plenty of towing ropes are loaded on motor transport, and horse-drawn and hand-drawn sleighs are considered very useful for transporting weapons and supplies.
(2) *Halts.* - In contrast with normal German practice in mountains, halts are short when the temperature is very low. Motor transport vehicles are placed radiator to radiator. Snow is cleared under the vehicles, and some sort of foundation is provided for the wheels.
(3) *Restrictions.* - The Germans limit the use of the tanks and motorized units when the temperature is lower than 5 degrees above zero (Fahrenheit). Motorcycles are considered useless when the snow is more than 8 inches deep. Snow is regarded as a tank obstacle when it is higher than the ground clearance of the tank's belly. German tractors can negotiate snow up to 1 foot in depth; at 1 foot the use of snow-clearing apparatus becomes necessary. At very low temperatures, gasoline consumption is reckoned at five times the normal rate. Snow deeper than 1 foot 4 inches is considered impassible for pack animals.

b. Weapons

The German Army warns its mountain troops that:
(1) Distances are usually underestimated in clear weather and overestimated in fog and mist.
(2) At low temperatures weapons often fire short at first.
(3) Ammunition expenditure tends to rise very sharply when visibility is bad.

c. Reconnaissance

When German reconnaissance units are operating in mountains under conditions of extreme cold, extra tasks include obtaining information about the depth of snow, the load capacity of ice surfaces, and the danger of landslides and avalanches.

Regarding direction signs, the Germans warn that the opposition will use every possible form of deception, and that great care must be taken in interpreting the direction of trails correctly. It is acknowledged that there is an ever-present danger of being diverted into an ambush or a strongly defended position.

The usual German methods of indicating trails are to mark trees and rocks, erect poles, and set up flags on staffs. Stakes are used to denote the shoulders of roads.

d. Signals

It is noted that a greater length of time is needed for laying communication wire under mountain conditions, and that cold and dampness lowers the efficiency of a great deal of signal equipment.

e. Attack Tactics

(1) Because of the difficulty of movement, assembly areas are near the opposition than is normally the case.
(2) Limited objectives are the rule.
(3) Because deployment is so difficult, it is often delayed until contact has been made.
(4) Combined front and flank attacks are used wherever possible.
(5) Commanding positions are considered of added value and are occupied by mobile troops as quickly as possible.
(6) Decentralization of weapons is authorized so that units can deal with surprise attacks without delay.
(7) Attacks are often made by ski troops.

f. Defense Tactics

German mountain troops are taught that under conditions of snow and extreme cold:
(1) Obstacles take much longer to build.
(2) Strong outposts are highly valuable because they force the opposition to undertake an early deployment.
(3) The usefulness of snow as protection against fire is often overestimated.
(4) Heavy snowfalls render mines useless.

Intelligence Bulletin Vol II, No 9. (May 1944), pages 61-65.

Section II. COMMANDERS, OBSERVERS DISCUSS ENEMY TACTICS

The following comments on German combat methods in Italy have been made by U.S. Army unit commanders and experienced observers. Only these tactics which have been noted repeatedly are mentioned here, since occasional, isolated instances of German methods cannot be regarded as illustrative of standard enemy procedure.

When the Germans suspect that new troops are opposing them, enemy patrols become very active, to determine the identity and strength of the new troops.

The Germans have used artillery and some rocket guns to harass our forward areas and to interdict vital supply roads, but fire has decreased as soon as the enemy has lost dominant observation.

The enemy often restricts his movements entirely to those he can make under cover of darkness or during days when weather makes hostile air activity impossible .... The Germans do not pull out in daylight, even when they have practically been surrounded. They fight like tigers to hold a narrow escape corridor, through which they try to withdraw at night.

Enemy agents make every effort to infiltrate into civilian traffic and movement.

Long-range weapons are active mainly on clear days. No change has been noted in the German policy of continually changing positions and of employing a considerable number of single guns.

It is necessary to stress again and again that road craters are surrounded and lined with antipersonnel mines -- often just over the lip of the crater, where they are harder to detect.

The Germans have used a large number of abatis in Italy, most of which have been invested with antipersonnel devices.

During the first part of December, the German artillery have the commanding observation, and, as we advanced, the enemy's activity was definitely in proportion to the visibility and observation advantages remaining in his hands. The enemy adjusted by observation the major portion of his artillery fire. Subsequent night harassing missions (German) were based on this data. German counterbattery fire decreased considerably whenever the advance of friendly troops deprived the enemy of commanding heights ... the Germans are continuing the policy of seeking to neutralize an installation temporarily, rather than exploiting opportunities for damage when an increased use of ammunition would be involved. Fire of medium caliber has been decreasing, with a corresponding increase in light caliber, principally 105mm guns.

The Germans will send out patrols to feign a night attack, or a daylight attack, just to locate your barrages so that they can side-step them when the real attack comes.

The German positions we have run into in the mountains have very few riflemen in the front line. The forward element of the defense has consisted almost entirely of machine guns in rock bunkers; these bunkers are so cleverly blended into the terrain that they are extremely difficult to locate. In the daytime the Germans seem to hold practically all their riflemen back about 200 yards. They depend on their machine guns, mortars, and artillery to stop your attack or to cause you such losses that a quick counterattack by the riflemen will throw you out. At night they put out listening posts manned by riflemen, but still hold back most of the riflemen.

On Hill 769 one of our companies got up close to the German bunkers. The company could not move in daylight because of the lack of cover, so a night attack was decided upon. Since there would be moonlight, it was decided to place smoke on the bunker at the time of the attack. This was done, but, as soon as the smoke screen was formed, the Germans left their bunkers, moved their right front and left front to the edge of the smoke screen nearest our positions, and placed machine-pistol fire on our attacking unit's flanks.

The outstanding feature of the mountainous country in Italy is that a village is almost invariably on the dominating ground, or on ground vital to the attacker to secure his line of communication. Such villages consist of closely packed houses with narrow streets between them. The houses themselves have thick walls and are immune to shell fire, except in the case of a direct hit. The enemy realizes this and makes full use of them as strong points, firing from windows and improvised loopholes. Such villages are also covered by machine-gun and mortar fire from either flank. In some cases houses are scattered on dominating features, and the enemy often uses them as machine-gun posts, covering the approaches by means of snipers and additional machine guns in adjacent houses.

In front of organized German positions, we have found mines only in the natural avenues of approach. These avenues are also covered with machine-gun and mortar fire. Thus the Germans are better prepared to deal with an opposing force using draws or gullies than one which is working its way along the sides of ridges. On the ridges and less jagged mountains, the Germans often dispose their strength on the reverse slopes in order to bring heavy fire on our forces as they cross over the crests.

Strong stone bunkers are continually being encountered in mountainous terrain. Although it is reported that grenades and rockets will not penetrate the walls of such bunkers, it has been found that both grenades and rockets are effective when exploded close to the slits, which are near ground level. The occupants are at least stunned. As a U.S. sergeant who has considerable experience with both weapons recently expressed it, 'If you close in fast after using them on bunkers, you will find the Germans either knocked cold or goofy." Another noncom observes, "Grenades exploding within 3 feet or so of the slit will get the Germans if they are looking out."
In town fighting, buildings and strong points occupied by the Germans have proven vulnerable both to the grenade and rocket launcher.

The German soldier does not like to fight at night, and does not fight as well at night as he does during the day. [ed. There is contradictory evidence on this point. See pages 38 & 72 .] In several instances German security at night has been found to be lacking. A number of instances have also shown that the German soldier, when surprised at night, has become confused and has been a easy victim of an opponent well trained in night fighting.

When patrols are sent out to locate the German defensive positions, (footnote: This commander is speaking of heavy stone bunkers.) the Germans do not fire on these patrols if they can avoid it, but let them go on through. German prisoners have stated that they were ordered not to fire except in the case of major attack. They have also recited instances of seeing our patrols go by their positions at a given time on a certain night. Checking back, we have found that our patrols were there at the stated time.

On the other hand, if one of our patrols stumbles into a German position, the Germans try to destroy the patrol to the last man, to keep the information from getting back to our units.

Repeatedly the entire patrol has returned to report that a hill is unoccupied or that a bridge has not been blown. Some unit moves forward and finds the hill alive with Germans, who smother the unit with fire from machine pistols, light machine guns, and mortars -- or, in the case of a bridge, the unit will find that the Germans have demolished it in the meantime. In other words, the Germans are quick to exploit the situation if an opposing force fails to seize and hold ground until stronger elements have been brought up to hold it in force.

Experience has shown that the Germans will almost invariably launch a counterattack to break up an attack made by small infantry units. You can expect such a counterattack, usually by 10 to 20 men, not more than after you get close to the German positions. They are usually well armed with light machine guns and machine pistols, and counterattack by fire and movement. They keep up a heavy fire while small details, even individuals, alternately push forward. The Germans almost always attack your flank. They seldom close with bayonet, but try to drive you out by fire.....

The Germans keep a sharp lookout for radio antennas, and shell every one they see.

The enemy is skillful at radio intercept, and tries to draw a great deal of information by inference. He notes the peculiarities of individual radio operators, which can easily become a dead give-away to the location of units.

Intelligence Bulletin Vol II, No 4. (December 1943), pages 54-56.

Section III. USE OF INFANTRY WEAPONS AGAINST PARACHUTISTS

1. INTRODUCTION

The German Army attaches great importance to the use of infantry weapons against parachutists. A German document acquired by the United Nations in Sicily discussed the technique of employing rifles and machine guns for this purpose. The following extracts from this document should be regarded as supplementary to a more general article, "Principals of Defense Against Airborne Troops," which appeared in the *Intelligence Bulletin*, Vol II, No.3.

2. THE DOCUMENT

a. General

German infantry units must at all times be prepared to meet surprise attacks by parachutists.

Hostile parachute troops jump from an altitude of 3,000 feet or more, drop about 1,000 feet, and then open their parachutes; or they may jump at an altitude of 400 feet, and open their parachutes after a drop of about 100 feet. One must reckon with a speed of fall from 16 to 20 feet per second. When the first of these procedures is followed, the parachutists, in landing, are dispersed over a large area. When a platoon of parachutists follows the second procedure, it attains in the air a lateral dispersion of from 425 to 750 yards, a depth of about 325 yards, and a difference of altitude of 50 to 65 feet between jumpers.

In employing infantry weapons against hostile parachute troops, German soldiers will fire only on the order of a responsible commander, such as a platoon commander or squad leader. The individual parachutists—not their parachutes—constitute the proper targets.

While a parachutist is landing, he may be attacked with every likelihood of success. At this time he must free himself from his parachute, and is helpless. If his weapons are dropped separately, he must recover them. This, too, will occupy him for a few moments.

All arms must participate in the task of crushing a parachute attack. Moreover, the employment of every form of ground defense for this purpose has the definite effect of breaking down the morale of the hostile force.

b. Use of the Rifle

Riflemen will fire on hostile parachutists as soon as the latter are within a range of about 425 yards. In a moderate wind, a rifleman will aim at the center of his target. In a strong wind he will lead the moving target according to firing rules. The rear sight will not be changed while fire is in progress. Riflemen will also fire on ammunition and weapons containers.

Standing or kneeling firing positions should be assumed. However, the situation may justify a prone position. Each rifleman will fire at the parachutist nearest him. When the parachutist jumps from an altitude of about 400 feet, the riflemen will not have time to fire more than five aimed rounds.

c. Use of the Machine Gun

Machine gunners will use ordinary ball ammunition against parachutists. It is advantageous to include armor–piercing tracer bullets in ammunition belts, in a proportion of 1 to 3.

Fire will be opened with the rear sight set according to the actual distance of the target. With reference to wind velocity, the rules for aiming are the same as those in subparagraph b. The rear sight will not be changed while firing is in progress.

The machine gun may be fired from a bipod, as a light gun, or from a tripod, as a heavy machine gun; however, surprise parachute attacks will generally compel a machine gunner to fire from the shoulder of another man.

If parachutists are dropped in front of a position, they are to be met with concentrated machine-gun fire. If they are dropped beyond the fire position on the flank, they are to be fired upon successively; that is, a machine gunner will fire on the nearest parachutist, and will then fire on any who remain in the line of sight. A volley of sweeping fire on scattered parachutists is a waste of ammunition, and is strictly forbidden.

Intelligence Bulletin Vol II, No 6. (February 1944), pages 68-70.

Section III. CONCENTRATING THE FIRE OF 81MM MORTARS

1. INTRODUCTION

Many U.S. junior officers and enlisted men who have fought the Germans in Tunisia and Italy have emphasized the necessity for a wider and better understanding of how the Germans use their infantry mortars against United Nations forces. For this reason the following enemy discussion of concentration of fire by German 81mm mortars should be of special interest to *Intelligence Bulletin* readers.

In connection with this article, reference should be made to "German Infantry Weapons" (M.I.D. Special Series No. 14) which contains descriptions of the German 50mm and 81mm (footnote: Although this is an 8.1cm mortar, it is called an 8cm by the Germans.)

2. ENEMY INSTRUCTIONS

a. General

The fire of one or two [81mm] mortar sections may be concentrated to achieve greater effectiveness against suitable targets. The fire unit is the section, even when two sections or a platoon are engaged. Throughout an action, platoon and section commanders must concentrate fire on the most important targets. When several appear at the same time, it may be more effective to engage them one by one, and with concentrated fire. Concentration of fire can be very effective in defense against such targets as observation posts, machine gun nests, and assembly areas.

Good intercommunication is essential for rapid concentration of fire. For a single section, this intercommunication can usually be accomplished by word of mouth; for two sections or a platoon, a telephone line will be necessary.

Targets must be indicated as quickly as possible. The methods employed are:

a. Indication on the ground (This is possible only for single sections, or if sections are close together.)

b. Fire by "voice control" section (The platoon commander establishes his observation post near a section which fires on the indicate target with a single mortar.)

c. Use of reference points.

d. Use of a plan with numbered targets.

Ranging is normally done by a single mortar firing on a registration point, to make the most of the element of surprise. A range finder is very helpful for this, and should be borrowed from a machine-gun platoon if necessary. On receiving the range, the other mortar in the section will correct it for position. Fire for effect will be undertaken only after this fire for preparation, except when engaging fleeting targets or targets of considerable size. Digging-in the base plate is of great importance, especially when mortars have not undertaken fire for preparation. The possibility of danger to our own troops from rounds falling short must be considered when firing mortars which have not undertaken fire for preparation.

b. By a Section

The section commander may either entrust detachment commanders with fire control or carry it out himself. In the former case, he indicates the target, or portion of the target, to detachment commanders, who carry out ranging individually and report when they are on the target. He then orders fire for effect according to the situation. In the latter case, he either ranges both mortars himself or ranges only one of them, the detachment commander ranging the second mortar while registration is proceeding. The mortar-position noncom determines the position correction and passes the result of the second mortar.

c. By Two Sections or a Platoon

The platoon commander establishes his observation post, and details the section which is to be near him to serve as the "voice control" section. Intercommunication with the other sections is arranged. Concentration of fire of sections must be regulated both as to space and time. Sections will be allotted portions of the target, and section commanders will further distribute the fire of individual mortars. The tactical situation may make it necessary for sections to range gradually and at varying intervals. When ranging has been completed, the platoon commander will order fire for effect. The order will be passed by line, by the fire of the "voice control" section, or fire may be arranged on a time basis. The platoon commander will observe each section's fire and report corrections, but section commanders must also observe and attempt to improve their fire independently. Concentration of fire of several sections is easier if the sections are sited as close together as possible. In this case it may be possible for the ranging to be carried out by a single mortar.

Intelligence Bulletin Vol II, No 5. (January 1944), pages 1-19.

Section I. HOW THE GERMAN ARMY
USES SMOKE IN COMBAT

1. INTRODUCTION

Up to a certain point, the Germans use smoke in combat much as we do. They believe that smoke should be supplementary to other weapons, rather than a weapon itself. Incidentally, the German Army makes more extensive use of smoke candles than does the U.S. Army. Although German doctrine covering the effect of weather, wind, and terrain on smoke screens is almost identical with our own, the enemy theory of ground attack under cover of area smoke, discussed in paragraph 3 of this section, differs from U.S. methods.

It should be noted that as of yet the Germans have not used white phosphorus smoke, which not only is a highly effective screening, but one which causes causalties, as well. Also, it must be remembered that if gas warfare should break out, the enemy may use screening smoke to mask barrages of poison gas. The purpose of such a procedure would be to force our troops to put on gas masks whenever smoke is used against them.

This section is based entirely on German Army documents, which the U.S. Army Chemical Warfare Service has translated and recommended for publication.

2. SMOKE SCREENS

a. In the Defense
The following examples of German tactical use of smoke are representative:

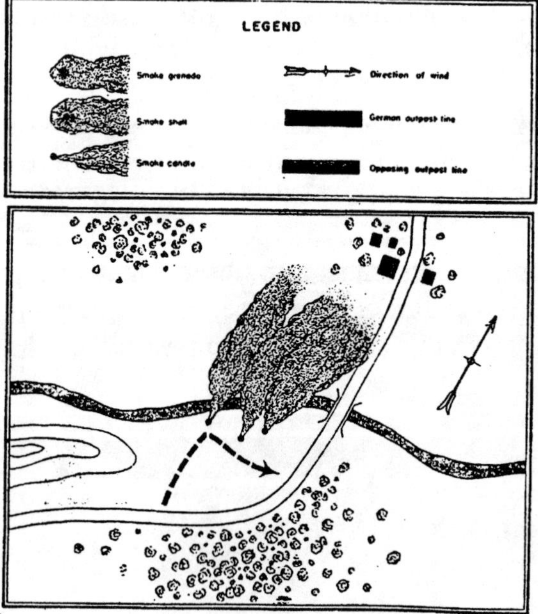

Figure 1.

(1) The advance guard in figure 1 must find out whether the group of houses in the upper right is occupied by United Nations soldiers. If it draws fire from these houses and from the grove of saplings at the upper left, smoke candles are ignited, and the advance guard returns to the woods under cover of the smoke.

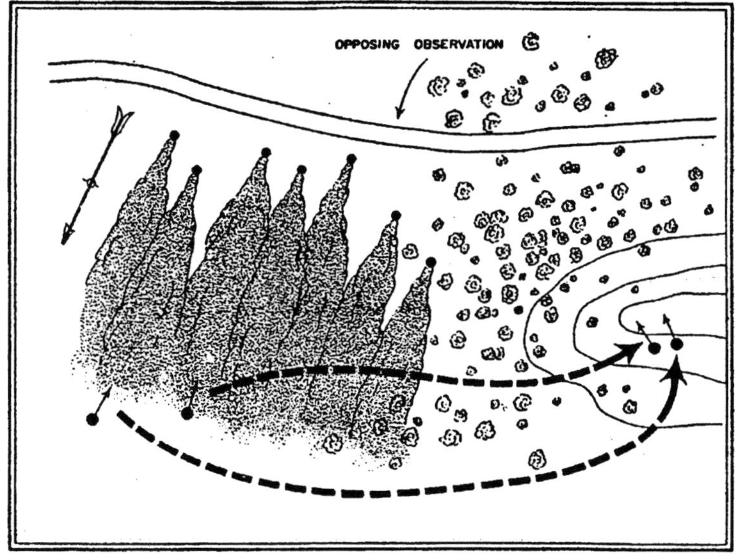

Figure 2.

(3) In figure 2 a German heavy machine gun platoon is under fire of United Nations artillery. With the wind coming from the 1:30-o'clock direction, the German platoon sets down a line of smoke candles, which permits movement to a new position on the hill at the right. It will be noted that smoke is used to cover only that part of the terrain which offers no concealment. The candles are of course placed with due regard for the direction of the wind.

(3) In figure 3 strong, well-spaced United Nations defenses have stopped a German attack at the entrance of a village. To prepare for further maneuvering later on, the Germans dig in under cover of smoke, taking advantage of all cover offered by the terrain.

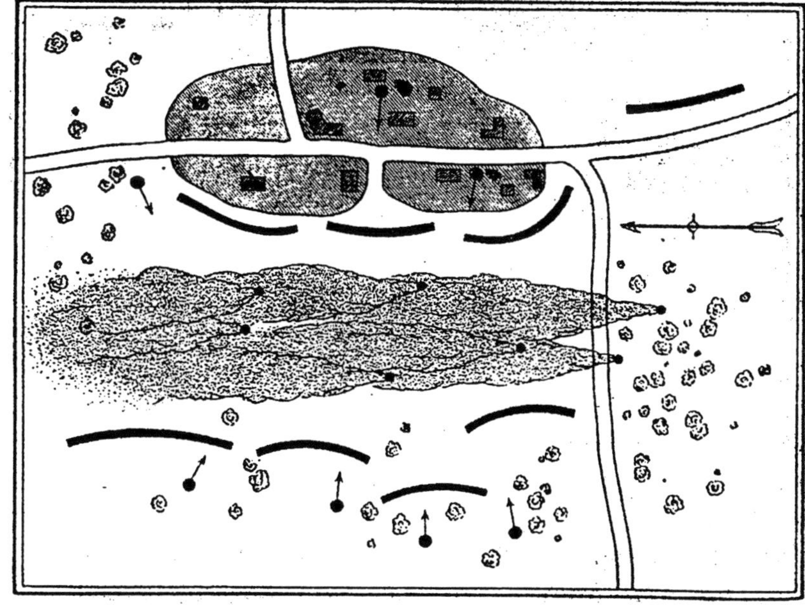

Figure 3.

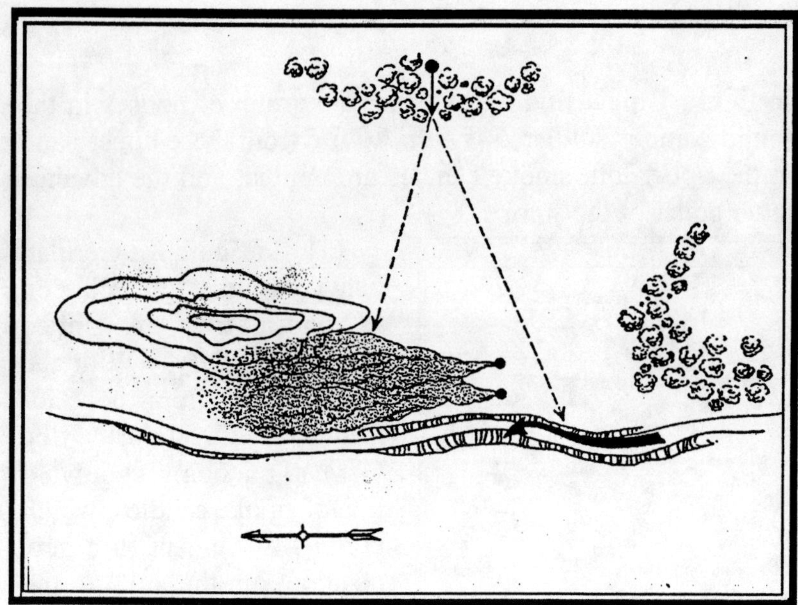

Figure 4.

(4) A German bicycle scout squad returning from reconnaissance suddenly receives flanking fire from the woods shown at the top of figure 4. The squad takes cover in a ditch, and ignites smoke candles. The smoke allows the Germans to proceed under cover behind the hill at the left. Behind this hill they are out of the field of fire.

(5) In figure 5 a German retrograde movement is taking place under cover of smoke. The withdrawal was begun as soon as the first screen was set up. Shortly afterward, the second screen was set up, to give the German unit time to reach the cover afforded by the woods shown at the bottom of figure 5. Plunging fire from machine guns on the flank is also covering the movement.

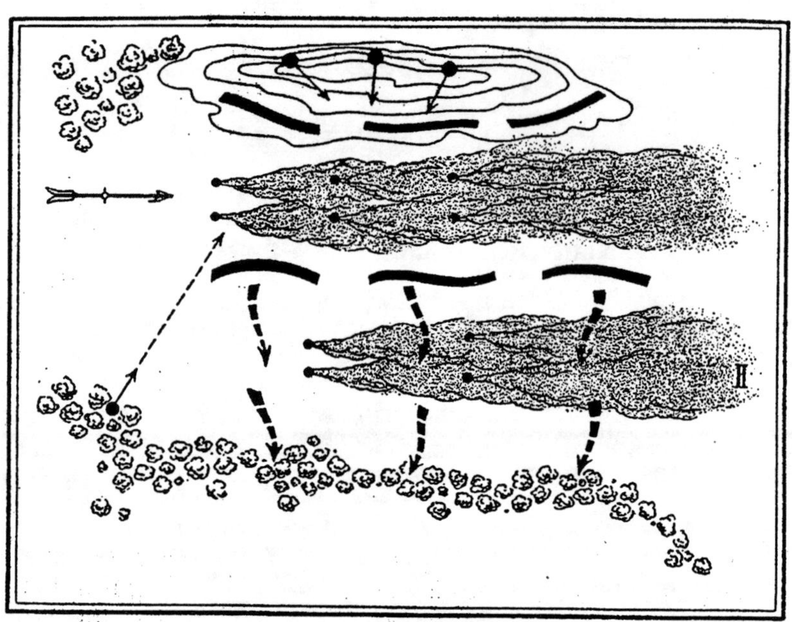

Figure 5.

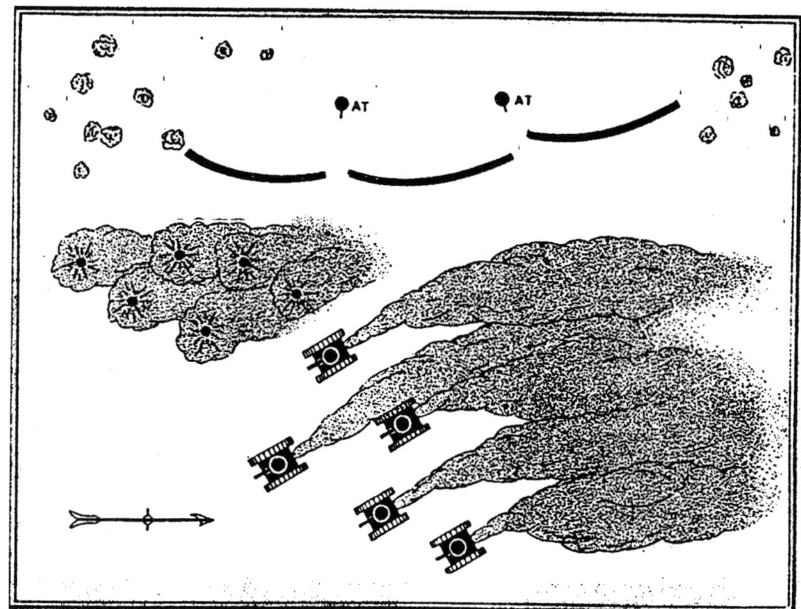

Figure 6.

(6) In figure 6 German tanks are withdrawing while screened by their own smoke. The flank vehicles are screened by the smoke from an artillery battery. It will be noted that this situation calls for quick realization by the artillery observation post of the tanks' predicament, and for a knowledge of how to use wind direction to achieve the proper secondary screen.

(7) *Antitank Defense*. -- In at least one instance during the North African campaign, a German tank unit made simple but effective use of smoke. Part of a United Nations armored division was brought up to help stem a German advance. It succeeded in ambushing a column of German tanks. Some damage was inflicted and the Germans withdrew, laying down a smoke screen. The commander of the United Nations force waited for the smoke to life, thinking it could not last long. But it persisted, and, since the terrain did not permit by-passing the screen, he gave orders for his tanks to proceed through it. As soon as the United Nations tanks were silhouetted on the other side, the Germans fired on them with everything they had, and inflicted great damage before retreating.

The Germans have also tried some interesting experiments to make tanks vulnerable. their glass smoke grenades seem to be designed specifically for antitank use, but might also be used against pillboxes. The Germans believe that by breaking smoke grenades over air-intake openings, it is sometimes possible to force a crew to evacuate its tank, or at least to have great trouble in handling it. Also, the Germans have experimented with the trick of tying smoke grenades -- probably with time fuzes -- to each end of a 6- or 7-foot rope and throwing it across the barrel of the tank's gun.

b. In the Attack

The following examples of German tactical use of smoke in the attack are representative:

(1) There is a barricade of wire obstacles in front of the United Nations machine-gun nest in figure 7. Under cover of smoke, the Germans cut gaps in the wire. (If there is any likelihood of prolonged activity in dense smoke, the Germans put on gas masks.) By means of a few smoke candles, the screen is extended toward the machine-gun positions, since the direction of the wind permits this. Both flanks of the next are attacked under the screen. Also, a few smoke grenades are thrown in front of, and into, the nest.

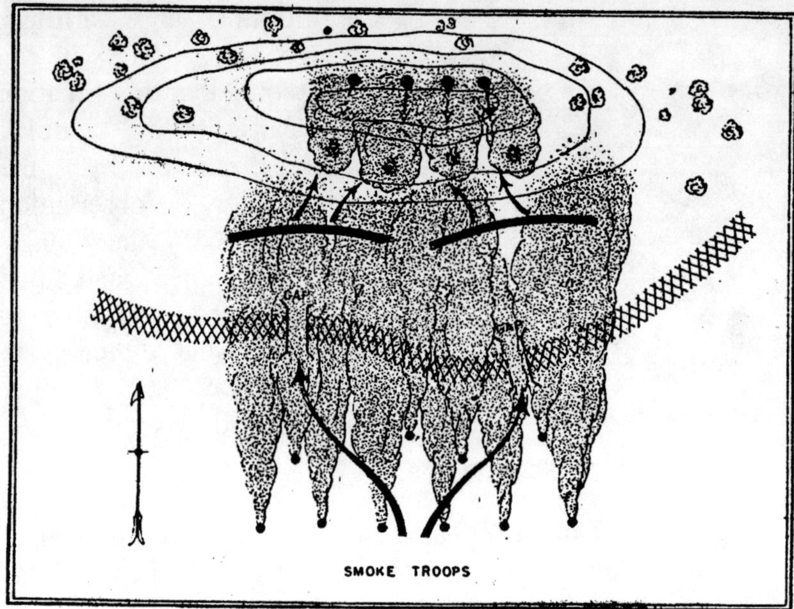

Figure 7.

For a large-scale attack on a United Nations line, the Germans are likely to follow a similar procedure. First, they may cover our centers of resistance with a heavy smoke blanket; then the attacking forces may try to envelop the positions from both flanks and reduce them from the rear.

(2) *Against Pillboxes* - The German technique of assaults against pillboxes is similar to that used against machine-gun positions [described in (1) above], with some interesting modifications. Usually the assault is preceded by concentrated artillery fire.

One purpose of this is to make craters in which an advancing combat engineer detachment can take cover. When the assault detachment reaches wire obstacles surrounding the pillbox, a Very signal calls for all available artillery fire to be placed on the pillbox. Smoke screens are laid down by grenades or candles.

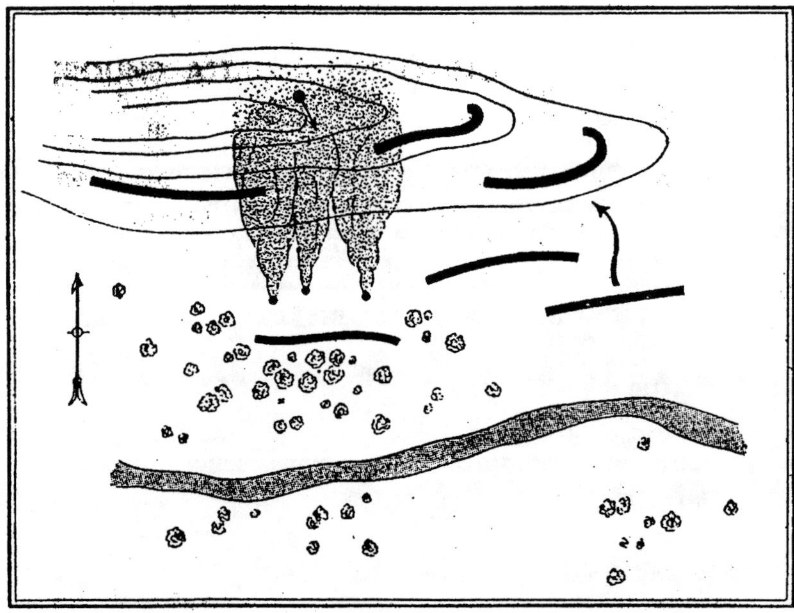

Figure 8.

Men with wire cutters or Bangalore torpedoes open gaps through the wire. Very signals call for artillery fire against the pillbox to cease, and a flame-throwing detachment advances through the gaps in the wire and tries to get within 5 or 6 yards of the pillbox. These men are covered by machine-gun fire. As soon as their fuel is almost exhausted, they shout a warning, and men with pile charges advance to the embrasures of the pillbox and detonate the charges inside it. If the pillbox still holes out, smoke candles may be thrown inside it to make the air unbreathable, or an attempt may be made to blow in the roof with a heavy charge.

(3) In figure 8 a German attack is stopped by fire from a heavy machine gun on the left bank. Smoke is used to blind the machine gun and the attack continues.

(4) During a German tank attack, a United Nations observation post and (suspected) antitank weapons (see fig. 9) are blinded by smoke from antiaircraft.

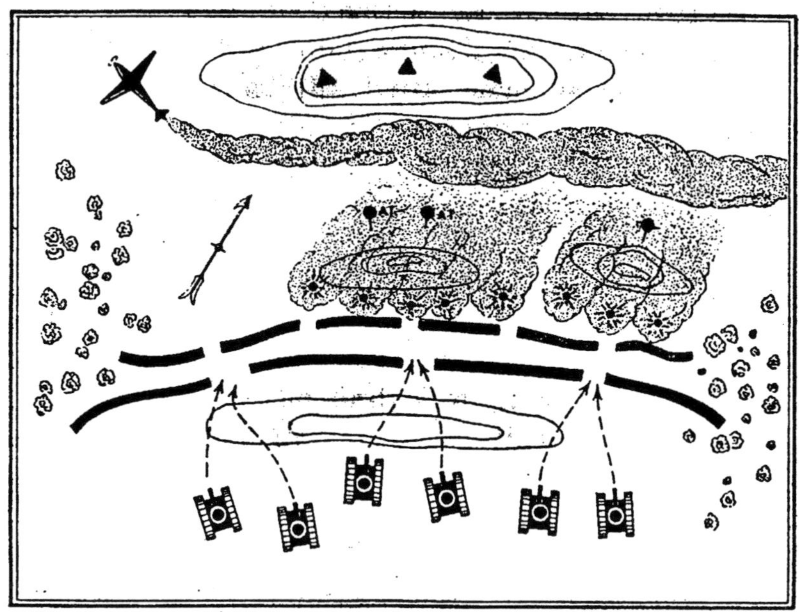

Figure 9.

3. GROUND ATTACKS EMPLOYING AREA SMOKE

a. Definition

The term "area smoke," as defined by a German manual, means the use of smoke to achieve, over an extensive area, an effect much like that of thick natural fog. The Germans regard this as an important asset in an attack against forces prepared for defense along a stabilized front, in a field defensive position, or behind a water obstacle.

Area smoke creates a zone in which, and into which, observation and observed fire are either difficult or impossible, except at close range. It therefore favors close combat, with infantry playing an important part. The Germans regard surprise as especially useful in an attack under area smoke; this prevents the opposition from rearranging its defensive strategy so as to safeguard against penetration of its lines.

b. Preparations

Before the main attack, the Germans occupy suitable positions of readiness. Reconnaissance is performed and thorough preparations are made, before the area smoke is laid down. The attacking units are given as clear a picture as possible of the nature of the terrain to be crossed, and details regarding the defensive capabilities of the opposing force.

The main objectives of the attack are the United Nations artillery positions. The Germans realize that if the main defensive zone is of great depth and strength, it may be necessary for them to decide upon intermediate objectives, which can be recognized easily in the smoke -- that is, roads running at right angles to the advance, intersection ofstreams, and so on. This makes it possible to correlate the laying of the screen with the progress of the attack.

Before the smoke screen is laid down, the artillery places thorough destructive fire on the area to be attacked. High ground, suitable for observation posts in the vicinity of the battle zone, is blinded. The German intention is to neutralize United Nations artillery while the attacking forces are moved toward their preliminary positions, and to keep it neutralized throughout the battle.

Minefields are cleared beforehand and obstacles are dealt with, so that the attacking forces will be delayed as little as possible. The night before the attack is to begin, the first wave of attacking units is brought up near the positions of readiness. Roads and paths approaching the defensive zones are so marked that, even after smoke has begun to drift over these routes, there can be no possibility of error. The attacking units come to the positions of readiness in open formation.

First light in the morning is regarded as the most favorable time for the attack to begin. The approach of the attacking units may be preceded by a fairly long area of smoke bombardment on the hostile forward positions. Smoke- and high-explosive projectiles are fired together until the attacking German infantry approaches the area which is being smoke-screened. If possible, the hostile area observation posts are blinded at this time.

c. Maintaining the Area Smoke Screen

The area smoke is laid down in zones 225 to 350 yards wide, across the axis of attack. Its rate of advance is governed by the difficulties that the Germans estimate they will encounter in the respective combat areas, and by the nature of the terrain; normally, the rate of advance averages about 225 yards in a 15-minute period. The Germans stipulate that once this screen has been set in motion, it must follow the timetable prepared in advance.

German units carry the attack forward under full cover of the smoke. At this stage, forward visibility is kept to hand-grenade range.

A head-on wind gives the attack the best smoke protection, the Germans believe. Also, a lateral wind has certain advantages, because it carries the smoke across sectors which are not being attacked -- a useful deception. The Germans say that a wind blowing in the direction of the United Nations forces is unsuitable, since it prevents use of the screen for close combat. The Germans believe in extending a screen over the flanks of the combat area, as a ruse. During an attack the screen is supplemented by the smoke equipment of individual attacking units -- that is, by smoke grenades and candles and by smoke shells from infantry guns. The Germans try to supply their units generously with these items, especially with smoke candles.

d. How Units are Coordinated

The division commander assigns objectives to the infantry commander, and also assigns missions to the artillery. Antiaircraft defense is provided while units are assembling. The area screen is laid down by smoke units and artillery, working together under the sole command of the artillery commander. This area screen is laid on the basis of a time schedule and a coverage outlined by the division commander.

The infantry regimental commander gives the battalions details regarding their attack and the objectives of the breakthrough. He puts under the command of the battalions such matériel as may be necessary to the fulfillment of their missions; this includes antitank guns, infantry guns, and aids to maintaining direction. Forces assigned to attack defensive works are augmented by combat engineer units with their special equipment, individual armored vehicles, and weapons for engaging pillboxes. The battalion commander goes forward with the troops engaged in the breakthrough.

Radio apparatus is considered necessary, even for lateral communication, during the course of fighting in the smoke-covered area. Additional signal units are attached, to provide the necessary communication under these difficult circumstances.

Reconnaissance aircraft are employed to report the effectiveness of the screen, and to reconnoiter the terrain beyond it. Dive-bombing attacks are carried out against hostile battery positions, assembly areas, and moving columns of hostile reserves.

After the breakthrough, the regimental commander reorganizes those forces which have become scattered.

3. The Attack

The infantry are the first to enter the combat area. The Germans are likely to dispatch assault detachments against individual and especially formidable defensive positions on the near edge of the combat area; the purpose of this, of course, is to make penetration easier.

German assault troops are thrown in at points where the terrain and the character of defensive obstacles offer conditions most favorable for a thrust. During the battle these assault troops must rely on their own resources. The German theory is that such units should be strong enough to fight their way across the combat area, and reach the far edge still strong enough to continue active fighting. It is a German principle that these assault units should advance simultaneously, no matter how great the variety of tasks that each unit must perform.

When the infantry is nearing the far edge of the hostile defenses, under cover of the smoke, the artillery begins its mission of providing the necessary fire cover for the spearhead of the attack, as it comes out of the smoke and within the view of United Nations forces. German artillery forward observers, with field telephone apparatus, accompany the forward elements of the attacking infantry, and communicate with the artillery liaison officers at battalion headquarters. During the advance through the smoke, the observers signal to field artillery observation personnel at prearranged times, to indicate which objectives have been attained. Vertical light signals, such as Very lights, are used for this purpose.

f. Aids to Maintaining Direction

The Germans consider it necessary for attacking units to be properly oriented at all times, so that there will be no confusion when vision is sharply reduced. The following devices are employed in maintaining direction:

(1) The magnetic compass.

(2) A radio beam set out by a transmitter, with several receivers picking it up. The transmitter is set up at the line of departure, and lays a beam bout 65 feet wide through the smoke, in the direction of the objective. The men operating the receivers can tell at any time whether they are in the radio beam or have deviated to a flank. A special unit of the signal service, moving with the leading elements of the attacking forces, operates this equipment.

(3) "Direction shells," which scatter red, yellow or blue powder, are issued to the infantry gun companies. Before the attack these rounds are so fired as to mark the path of the attack with colored patches every 55 yards (approximately).

(4) Marking lines, each about 350 yards long, attached to rockets.

(5) The gyro-compass.

(6) Direction tapes in different colors mark the routes that staffs or units have taken through the smoke. These tapes facilitate courier traffic, the maintenance of contact, and the forward movement of units advancing later on. Numerals on the tapes tell men fighting within the screen how far they have penetrated from the line of departure.

g. Follow-up of Attack

When the attacking units reach the far edge of the screen, they recognize so that an immediate attack can be launched against important objectives, which usually lie beyond the screened area. While the infantry is reorganizing, antitank guns provide protection. Also, tank units which have been held in readiness are brought through the newly captured area to attack hostile artillery positions and any hostile reserves may have been brought up. German reserves exploit the breakthrough so that maximum strength can be applied against United Nations forces on the far side of the smoke screen.

h. Note on Area Screens Used with Water Obstacles

It is a German principal that, when hostile forces are defending themselves behind a water obstacle, an area screen is useful only if there is little or no current or if the stream is a narrow one. The first wave to be ferried over consists only of assault troops who are to engage the fieldworks on the opposite bank. The timetable for the area screen takes into account the requirement that units ferried across the stream must have time to assemble on the opposite bank before any thrust against extensive hostile defenses is attempted.

Section VI. COMBAT SMOKE SCREENS RECENTLY USED IN ITALY

The Germans have been using smoke on an increased scale during recent operations in the Cassino and Anzio beachhead areas. In the beachhead fighting, especially, smoke was used on a large scale on a number of occasions.

In *Intelligence Bulletin* Vol II, No. 5, pp 1-19 [see preceding article in this document], the German theory of smoke tactics was outlined in some detail. The following report from a U.S. observer in Italy gives actual examples of the use of smoke, both in the attack and in the defense:

On Monday afternoon, 28 February, the Germans set up a smoke screen about 4 miles long, and maintained it from 1630 until after dark. The wind was favorable, and under cover of this screen, which was blowing parallel to our lines, the Germans re-aligned many of their units, and got their artillery in travel positions preparatory to making a push the following morning.

On Tuesday morning, the 29th, I witnessed the use of a smoke screen to cover a German attack on the Anzio beachhead. The smoke was laid with a 12-o'clock wind blowing in the face of the attackers. The screen was placed well back of our front lines, on our main support position. In the cool damp air of the early morning, this smoke cloud settled down to a solid bank, which moved across the level fields and passed over our front lines. However, it was fairly well dissipated by the time it reached the Germans.

From their high observation points in the mountains to the rear, the Germans were able to see over this cloud and to direct their artillery fire against specific targets, while at the same time, the view of our observers was cut off over that entire front. The Germans made a considerable dent in our lines; however, this was more than straightened out that night, when our infantry counterattacked under cover of our own smoke screen and air bombardment.

German platoons and detachments attempted to infiltrate into our lines. When counterattacked, the Germans usually set up a smoke screen with hand grenades and small smoke pots in an attempt to cover a withdrawal.

German tanks invariably use their smoke-screen apparatus -- that is, their smoke projectors -- when they are fired upon. The tanks then move to safer places under cover of a smoke screen.

When small German units are preparing a night attack, they almost invariably set up a smoke screen about half an hour before darkness, and, behind the screen, move into their new attack positions. Also, smoke screens often are set up when no attack is intended. This is done with the idea of harassing our front-line units into making defensive preparations which involve a waste of time and energy.

Registration is done in the early morning and late afternoon -- very frequently with smoke shells, and sometimes with only two or three rounds. On the Anzio beachhead, following the attack on the 29th, it was quite noticeable that the Germans were registering with about three rounds of smoke on all the cross roads and the various draws that our troops might conceivably use in a night movement.

About 90 percent of the German smoke shells now being used are believed to be filled with a brown-tinted liquid which gives off a dense white smoke. About 10 percent, which are used for harassing purposes, rather than for screening, are filled with white phosphorus. The use of white phosphorus by the Germans began about three months ago, and has gradually been increasing. Most of the smoke produced within the Germans' own lines apparently is created by smoke pots.

The foregoing methods of employing smoke appear to be practically standard, inasmuch as they have been used in exactly the same way on the Anzio beachhead and on the Cassino front.

Intelligence Bulletin Vol II, No 3. (November 1943), pages 9-15.

Section II. SIX-BARREL ROCKET WEAPON (THE "NEBELWERFER 41")

1. INTRODUCTION

Whenever the fortunes of the German army take a new turn for the worse, Nazi propagandists attempt to encourage the people of the Reich -- and influence public opinion in neutral companies -- by spreading rumors of new and formidable developments in German ordnance. Recently the Nazis have been releasing propaganda declaring that spectacular results are being achieved with the German six-barrel rocket projector known as the *Nebelwerfer* (smoke mortar) *41*. Actually, this is not a particularly new weapon. Its name, moreover, is extremely misleading. In the first place, the *Nebelwerfer 41* is not a mortar at all, and, in the second place, it can accommodate both gas-charged and high-explosive projectiles, as well as smoke projectiles.

It would be just as foolish to discount the German claims 100 percent as it would be to accept them unreservedly. Although fire from the *Nebelwerfer 41* is relatively inaccurate, one of the weapon's chief assets appears to be the concussion effect of its high-explosive projectiles, which is considerable where the weapon's six barrels are fired successive, 1 second apart. The high-explosive round contains 5 pounds of explosive; this is comparable -- in weight, at least -- to the high-explosive round used in the U.S. 105mm howitzer.

In view of the mass of misleading information which has been circulated regarding the *Nebelwerfer 41* -- or as the Germans sometimes call it, the *Do-Gerät* -- it is hoped that junior officers and enlisted men will find the following discussion both timely and profitable. (footnote: U.S. soldiers in Sicily promptly nicknamed the *Nebelwerfer 41* the "Screaming Mimi.")

2. DESCRIPTION

The *Nebelwerfer 41* (see figs 1 and 2) is a six-barreled (nonrotating) tubular projector, with barrels 3 to 3 1/2 feet long and 160mm in diameter. The projector is mounted on a rubber-tired artillery chassis with a split trail.

There is no rifling; the projectors are guided by three rails, each about 1/3-inch high, which run down the inside of the barrels. This reduces the caliber to approximately 150mm.

The barrels are open-breeched, and the propellant is slow-burning black powder (14 pounds set behind the nose cap). This propellant generates gas through 26 jets set at an angle. As a result, the projectiles rotate and travel at an ever-increasing speed, starting with the rocket blast. The buster, which is in the rear two-sevenths of the projectile, has its own time fuze. The range is said to be about 7,760 yards.

Figure 1. German Six-barrel Rocket Projector (side view)

The barrels are fired electrically, from a distance. They are never fired simultaneously, since the blast from six rockets at once undoubtedly would capsize the weapon. The order of fire is fixed at 1-4-6-2-3-5.

The sighting and elevating mechanisms are located on the left-hand side of the barrels, immediately over the wheel, and are protected by a light-metal hinged box cover, which is raised when the weapon is to be used.

Figure 2.—
German Six-barrel Rocket Projector (front view).

Each barrel has a metal hook at the breech to hold the projectile in place, and a sparking device to ignite the rocket charge. This sparker can be turned to one side to permit loading and when turned back so that the "spark jump" is directed to an electrical igniter placed in one of 24 rocket blast openings located on the projectile, about one-third of the way up from the base.

About one-third of the length of the projectile extends below the breech of the weapon.

The projectile itself resembles a small torpedo -- without propeller or tail fins. The base is flat, with slightly rounded edges. The rocket jets are located about one-third of the way up the projectile from the base, and encircle the casing. The jets are at an angle with the axis of the projectile so as to impact rotation in flight, in "turbine" fashion.

The propelling charge is housed in the forward part of the rocket. A detonating fuze is located in the base of the projectile to detonate the high-explosive or smoke charge. In this way, on impact, the smoke or high explosive is set off above ground when the nose of the projectile penetrates the soil.

3. NOTE ON OPERATION

The following note on the operation of the *Nebelwerfer 41* is reproduced from the German Army periodical *Die Wehrmacht*. It is believed to be substantially correct.

"The *Nebelwerfer 41* or *Do-Gerät*, is unlimbered and placed in position by its crew of four men. As soon as the projectile coverings have been removed, the projector is ready to be aimed and loaded. The ammunition is attached to the right and to the left of the projector, within easy reach, and the shells are introduced two at a time, beginning with the lower barrels, and continuing upward. Meanwhile, foxholes deep enough to conceal a man in standing position have been dug about 10 to 15 yards to the side and rear of the projector. The gunners remain in these foxholes while the weapon is being fired by electrical ignition. Within 10 seconds a battery can fire 36 projectiles. These make a droning pipe-organ sound as they leave the barrels, and, while in flight, leave a trail of smoke. After a salvo has been fired, the crew quickly returns to its projectors and reloads them. "

Intelligence Bulletin Vol II, No 12. (august 1944), pages 54-55.

SMOKE-SHELL TACTICS USED BY GERMAN TANKS

As a rule German tanks employ smoke shells to achieve surprise, to conceal a change of direction, and to cover their withdrawal. The shells normally are fired to land about 100 yards in front of an Allied force. There are no reports to indicate that smoke shells are used in range estimation.

In attacking a village, German tanks fire smoke shells to lay a screen around the village in an effort to confuse the defenders as to the direction of the attack. Smoke shells always are used to conceal a change of direction of the attack, the wind permitting. When a German tank company (22 tanks) wishes to change direction, smoke shells are fired only by one platoon, with the fire tanks of a platoon firing three shells and high-explosive shells. If the force does not know the exact location, only smoke shells are used. When a single tank runs into an antitank position, it likewise fires only smoke shells, usually two or three rounds, to cover its movements.

Smoke shells are fired from the 75mm guns of the Pz.Kpfw. IV's, and also, it is reported, from 88mm guns on other armored vehicles. Smoke shells are not fired by the Pz.Kpfw. II or the Pz.Kpfw. III, both of which are equipped to discharge "smoke pots" with a range of approximately 50 yards. These pots are released electrically, and are employed chiefly to permit the tank to escape when caught by antitank fire.

Intelligence Bulletin Vol II, No 3. (November 1943), pages 26-36.

<u>Section V. TACTICAL EMPLOYMENT OF FLAK IN THE FIELD</u>

1. INTRODUCTION

The original German doctrine regarding the employment of German Air Force flak artillery in the field has steadily been undergoing modification. German manuals formerly described the responsibility of flak in the field as primarily, and almost exclusively, antiaircraft defense; the engagement of ground targets was regarded as secondary, and only to be undertaken in an emergency. Although the older manuals admitted the possibility of using light flak to reinforce the fire of heavy infantry weapons, and of using heavy flak to supplement antitank and other artillery, such employment was described as exceptional. There was nothing to suggest, for example, the now extensive use of the 88mm antiaircraft gun in an antitank role.

The transition from the defensive doctrine of the earlier manuals to the more aggressive modern conception seems to date from the introduction of the Flak Corps – units of which first appeared during the Battle of France. The Flak Corps was created to perform the tasks described in the following enemy notes.

"The Flak Corps is a wartime organization, and constitutes an operational reserve of the commander in chief of the German Air Force. It combines great mobility with heavy firepower. It can be employed in conjunction with spearheads composed of armored and motorized forces, and with non-motorized troops in forcing river crossings and attacking fortified positions. It can also be deployed as highly mobile artillery to support tank attacks."

"The Flak Corps can take part in antitank defense on a broad front, and can be employed in ground engagements at strongly contested points. Its capabilities are tremendous in antiaircraft defense, because its great mobility enables it to rush flak concentrations to strategically important points, and to transfer flak strength from one area to another, as required."

"It is also responsible for protecting forward ground organizations of the German Air Force."

As these notes show, flak in the field is not intended to serve as a powerful and highly mobile striking force. The emphasis laid on its employment in the ground role, and in an offensive capacity in conjunction with spearhead formations, is most important. Experience has verified that these principles are actively practiced in the field.

2. HEAVY FLAK
a. General

In operations with the field army, the 88mm gun, as a result of its great mobility, has become almost the universal weapon of heavy flak. Larger calibers are usually encountered only in areas where the defense is static.

The heavy flak battery consists of either four for six guns (usually 88's), with two light guns (20mm) for close protection. Six-gun batteries are becoming increasingly common. In theory the heavy battery consists of two platoons, but in practice it is rarely divided in this manner. All the guns are generally fitted with shields, to protect the detachments against small-arms fire, and with two sights – a telescopic sight for the direct engagement of ground targets, and a panoramic sight for indirect laying. In the interests of mobility, the fire-control equipment is often left behind. In addition to time-fuze high-explosive ammunition, armor-piercing ammunition, armor-piercing, and percussion-fuze high-explosive ammunition is normally carried. To avoid the muzzle flashes which, at night, readily give away gun positions, the Germans now make widespread use of flashless propellant.

The 88mm gun can be put into action in about 2 minutes. If necessary, it can be fired from its mount, but against ground targets only.[5] Since the normal mount is conspicuous because of its height, the gun is extremely vulnerable to artillery fire. Whenever possible, therefore, the gun is dug in so that only the barrel appears over the top of the emplacement. (Actually, the time factor and frequent moves do not always permit the Germans to devise effective concealment.) Realizing that destruction of hostile observation posts constitutes an indirect method of protecting their heavy flank guns, the Germans try to accomplish this at every opportunity.

The 88mm guns can open fire on armored vehicles at 2,500 yards with fair prospect of success, but are most effective at ranges of about 1,000 to 1,500 yards. They may fire at ranges of as much as 4,000 yards, if other and more inviting targets are not available. With the aid of a forward observation post, 88's sometimes engage such targets as troop concentrations at ranges of as much as 6,000 yards.[6] The following are

---

[5] Against ground targets on the Eastern Front, the Germans have used a self-propelled 88mm gun, called the "Ferdinand."

[6] The telescopic sight is graduated up to 10,340 yards, and theoretically it would be possible to engage targets up to this range. In indirect fire, when the panoramic sight is used, the maximum range of the 88mm Flak 36 is 11,45 yards with time-fuze high-explosive ammunition, and 16,132 yards with percussion-fuze high-explosive ammunition.

Corresponding maximum ranges with the 88m Flak 41 are:
Direct fire (with telescopic site) -- 11.770 yards
Indirect fire (with panoramic sight):
    Using time-fuze HE -- 13,561
    Using percussion-fuze HE --- 22.091

examples of the penetration performance with the 88mm Flak 36, the most common model of this gun:

Range (Yards)	Thickness of Armor	
	30° angle of impact	Perpendicular (no angle of impact)
500 -------------------------	110mm (4.33 in) -------------	129mm (5.07 in).
1,000 -------------------------	101mm (3.97 in) -------------	119mm (4.68 in).
1,500 -------------------------	92mm (3.62 in)---------------	110mm (4.33 in).
2,000 -----------------------	84mm (3.30 in) -------------	100mm (3.93 in)

It is estimated that the following figures are correct for the 88mm Flak 41:

Range (Yards)	Thickness of Armor	
	30° angle of impact	Perpendicular (no angle of impact)
500 -------------------------	150mm (5.91 in) -------------	175mm (6.89 in).
1,000 -------------------------	140mm (5.51 in) -------------	164mm (6.46 in).
1,500 -------------------------	130mm (5.12 in)-------------	153mm (6.02 in).
2,000 -----------------------	121mm (4.76 in) -------------	142mm (5.59 in)

b. Employment in Rear Areas

In rear areas heavy flak has the normal task of providing antiaircraft protection for ports, airfields, dumps, headquarters, and points of importance on lines of communications. Predictors and/or auxiliary predictors are employed, and mobile radio-location equipment may be allotted. Although flak units in rear areas primarily have the task of providing antiaircraft protection, even these units are normally provided with armor-piercing and percussion-fuze high-explosive ammunition, and therefore can operate against any hostile troops or armored vehicles which may break through. The heavy flak's degree of preparedness to meet such attacks naturally depends on the distance between the guns and the front.

c. Employment in Forward Areas

It is in the employment of heavy flak batteries attached to the Army, for operations in forward areas, that current German methods depart most noticeably from the doctrine expressed in earlier manuals. Formerly, German doctrine outlined a primary antiaircraft role, a secondary antitank role, and, under exceptional circumstances, employment in a field-artillery role. It may be said that the antitank role now has assumed virtual priority, for experience has shown that the 88mm gun has become an indispensable complement to the German Army's antitank artillery. A certain proportion of heavy batteries in forward areas is still deployed in an antiaircraft role, chiefly to protect forward airfields, and during periods of inactivity or preparation the antiaircraft role still predominates. For example, an assembly prior to an attack will usually be protected by heavy guns, and under these conditions the ground role is assumed only in the event that the Germans are subjected to a surprise attack. However, once battle is joined, whether in attack or defense (and especially when armored forces are involved), the heavy flak guns are usually employed against ground targets only, and the antiaircraft role becomes the exception. If necessary, even guns originally deployed to give antiaircraft protection to forward airfields are sometimes pressed into service as antitank weapons.

The employment of heavy flak batteries naturally varies considerably, depending on the terrain and the nature of the fighting. In open country the 88mm gun's long range gives it a distinct advantage as an antitank weapon. In North Africa, where so often there was no well-defined "line," heavy flak batteries often served as the nucleus of defensive "hedgehogs." In an advance the primary function of the batteries usually has been to provide antitank protection during the movement of German armored vehicles. The 88's have also been known to accompany tanks in an assault-gun role. Although the battery is the normal fire unit, large numbers of 88mm guns have occasionally been employed under one command when the situation has required that maximum antitank strength be concentrated at a single point.

A striking example of the value of heavy flak in defense is afforded by the final phases of the Tunisian campaign, in which heavy flak units frequently provided the backbone of German resistance to the Allied advance. For this purpose several units were formed into mobile battle groups, a procedure which had been reported to on previous occasions, and which presumably is dictated by the stress of circumstances. These flak battle groups are purely temporary units, formed for a specific purpose. They consist of a number of platoons, usually with two heavy and three light guns each, and may be employed either alone or in combination with other arms. They are used both in defense and attack. Since they are mobile striking forces, there is always a possibility that they will be used by the Germans in attempts to repel landings on the European continent. They would afford a means of rapid counterattack in threatened sectors. The employment of these temporary units, which has become increasingly common, demonstrates the flexibility of flak organization in the field and the extent to which the Germans use heavy flak to complement antitank artillery.

The employment of a heavy flak battery is naturally governed by the type of operation that is being undertaken by the Army unit to which it is attached. Although the lessons learned from desert warfare are not necessarily applicable to other theaters, the activity of a heavy flak battery during the early stages of the German counteroffensive in Cyrenaica in May 1942 affords some very good tactical illustrations. During this action the battery accompanied the Army unit to which it was attached, and provided protection both against air attacks and ground targets.   The ground role predominated. Not only were tank engagements fought by day, but at night the battery was deployed in an antitank role to protect its "parent" Army unit. The battery was continually on the move during the day. More than once it detached some of its guns to strengthen another Army unit, and at other times it, in turn, was given added strength. When opposition was expected, the battery took up an antitank siting, generally on high ground and facing the probable line of attack. The choice of this position was not hard-and-fast. The battery moved to a different position when reconnaissance had established the location and course of the hostile tanks. When in position, the battery often had to site its guns so that they faced in two directions, because of uncertainty as to the exact line of attack.

3. LIGHT FLAK
a. General

Light flank units operating in the field are generally equipped with 20mm guns (single- or four-barreled), sometimes with 37mm (1.45 in.) guns and once in a great while with 50mm (1.97 in.) guns. A light battery normally consists of four platoons of 20mm guns or three platoons of the larger caliber guns, with three guns to each platoon.

Light flak guns are especially useful in combating surprise attacks, because of the rapidity with which these pieces can be put into action. The 20mm Flak 30, for example, can be put into action in about half a minute, and in extreme emergencies all light flak guns can be fired (although with a limited traverse) from their mounts. In addition, it is known that self-propelled models of the 20mm and 37mm calibers exist and can engage both air and ground targets. Like the heavy guns, the light guns in the field are usually fitted with shields for protection against small-arms fire. They are also fitted with flak sights and/or telescopic or linear sights, and carry armor-piercing ammunition in addition to percussion-fuze high-explosive ammunition. Light flak guns may engage ground targets, especially "soft-skinned" vehicles, at ranges of as much as 800 yards, but are most effective at ranges up to about 300 yards. The following are examples of the penetration performance of the 20mm Flak 30 firing armor-piercing projections:

	Thickness of Armor	
Range (Yards)	30° angle of impact	Perpendicular
100 -------------------------	31mm (1.22 in) -------------	48mm (1.89 in).
200 -------------------------	29mm (1.14 in) -------------	44mm (1.73 in).
300 -------------------------	27mm (1.06 in)---------------	41mm (1.61 in).
400 -------------------------	25mm (0.98 in) --------------	38mm (1.50 in)

b. Employment in Rear Areas

In rear areas light flak batteries have the normal task of giving antiaircraft protection to such vital points as airfields, bridges, railroad stations and junctions, headquarters and depots. For this purpose batteries are generally deployed as a whole, with the guns sited by platoons. Although the antiaircraft role predominates, these batteries constitute an important element in the ground defense plan for the vital rear points they are protecting, and are prepared to engage any armored or other forces which may succeed in penetrating to that depth.

c. Employment in Forward Areas

Light batteries attached to Army units in forward areas may also operate as a whole, but platoons are usually detached to perform special tasks.

On the march, platoons are generally spaced at intervals along the column, or are sited at particularly vulnerable points along the route – such as bridges, defiles, or crossroads – and subsequently "leapfrog" forward. Their principal task is to protect the column against attack by low-flying aircraft; their secondary task is to engage ground forces.

In battle light flak units afford protection for headquarters, field artillery concentrations, infantry concentrations, engineer units, motor parks, and so on. Also, it is sometimes considered to assign a light platoon (three guns) to a heavy flak battery engaged in antitank work – presumably because, under certain circumstances, the two light guns belonging to the battery do not afford enough protection. In all these tasks the antiaircraft role predominates, but engagement of personnel and armored vehicles is also regarded as highly important and often takes place. Experience has shown that during tank attacks, light guns, as well as heavy guns, have ignored air targets and have concentrated on hostile armored vehicles, leaving German ground units to defend themselves against air attack by means of rifle and light machine-gun fire. (As previous issues of the *Intelligence Bulletins* have explained, German Army trainings stresses the importance of small-arms fire in defense against low-flying aircraft.)

It will be seen that whereas heavy flak- which is well suited to combat ground targets, partly because of its penetration performance – is now being given wide tactical employment in a ground role, light flak with the Army still clings pretty muich to the principles outlined in German pre-war manuals. Although the capability of light flak in a ground role is always something to take into account, this type of employment seems to be the exception, rather than the rule.

Note: This section has dealt solely with German Air Force flak. There are also (1) Army flak (*Herresflak*) units, which include "mixed" Battalions (containing both heavy and light batteries) as part of the artillery, and (2) light companies (*fla),* which have light guns only, as part of the infantry. These other troops are not numerous, however. As a rule, they are GHQ troops, and are attached to army units in much the same way that German Air Force flak units are attached. Recently enemy documents show that an Army flak battalion, consisting of two heavy batteries and one light battery, is now included in the tables of organization of armored and motorized divisions.

Intelligence Bulletin Vol II, No 9. (May 1944), pages 26-36.

Section III. GERMAN MINE WARFARE AND BOOBY TRAPS (ITALY)

1. GENERAL

The terrain fought over in Italy has been especially well suited to the employment of mines as delaying and casualty weapons. The limited road net; the mountainous, close country, with few trains and secondary roads; the terrain conformation and nature of available approaches to objectives, and the numerous bridge sites and by-passes – all these have presented good opportunities for the employment of various types of mines. The element of rapid withdrawal and pursuit which characterized the fighting in Sicily has not been present in the Italian campaign thus far. For this reason the laying of mines and the preparation of booby traps has not been hurried. It has been deliberate, thorough, and widespread.

The general pattern of minelaying in Italy has fairly well paralleled that encountered in Sicily. The nature of the terrain has generally prevented the use of extensive minefields, such as those encountered in southern Tunisia.   As in Sicily, the most heavily mined areas have been roads, valleys, natural approaches to objectives, trails, suitable bivouac areas, and demolitions. Abandoned towns and villages have been thickly mined and booby-trapped. In his withdrawal, the enemy has had the advantage of knowing the most likely sites for artillery positions, and these have nearly always been thickly strewn with mines and booby traps. There have been no mined dry river and stream beds to cross, as was the case in Sicily. On the other hand, mined river banks have had to be negotiated in the several river crossing operations during the campaign.

2. TYPES OF MINES ENCOUNTERED

A number of different types of mines have been encountered in Italy, and some of these have been new. The different models of the Tellermine have been freely used by the enemy, and there has been an increased use of wooden mines, both of German and Italian design. In recent phases of the campaign, the wooden-type mines have largely exceeded the number of standard metal Tellermines used. A high-ranking U.S. Engineer Officer has made the following comments on the mines encountered:

> ... The Germans are now using mainly wooden box and S-mines. S-mines have been encountered lately in quantities never seen before. We are now running into increasing numbers of Italian wooden box mines equipped with German bakelite igniters set for pressure, pull, and release. They are often in fields, but not necessarily scattered at random. For example, a whole olive grove will be mined and booby trapped with S-mines or wooden box mines, or both. Since the landing at Salerno, about 20 percent of all mines encountered have been of the wooden-box type.[7] The German wooden box type is about 12 inches square by 5 inches thick, and contains about 10 pounds of Triton or similar explosive. The Italian ones are not quite so big.

In some areas newer types of improvised concrete antipersonnel mines have been discovered. One type is a spherical concrete case, 10 inches in diameter, enclosing standard German and Italian explosive charges equipped with standard types of igniters. Shrapnel has been used as aggregate in these concrete mines. Another type is a spherical concrete mine, 13 inches in diameter, cast in two halves, which are bolted together with steel rods. These have contained about 9 pounds of explosive and have been equipped to receive detonating devices.

3. NOTES ON NEW ENEMY METHODS
a. Smoke –Warning Booby Trap

The Germans are using a new type of booby trap, which consists of a German smoke canister, an incendiary detonator, a ZZ35 pull-igniter, and a trip wire.

The smoke canister is a cylinder 3½ inches in diameter and 5½ inches high, which is painted green with two white bands around the outside. It is marked "Nb. K-S-39B."

---

[7] In late January [1944] it was reported that wooden mines represented about 40 percent of all mines cleared during that month.

A threaded hole in the top will take a detonator and ordinary German booby-trap mechanisms. The detonators used are of a special incendiary type, and may be identified by a green band, ¾-inch wide, around the closed end.

This booby trap has no morale or casualty effect. However, if it is tripped during daylight hours, it will reveal the fact that movement is taking place. Presumably the German theory is that while the detonation of an S-mine or some other explosive charge may pass unnoticed by the defenders if considerable firing is in progress, the smoke from the booby trap will provide an unmistakable visual warning.

b. New Trip-wire Arrangement

Recently a new trip-wire system has been employed by the enemy. The trip wires are so arranged that German soldiers may trip over the wires without suffering casualties, whereas if advancing British or American soldiers trip over the wire, the detonation of the firing device may cause casualties to other men advancing behind them. This new method of arranging trip wires is illustrated in figure 21.

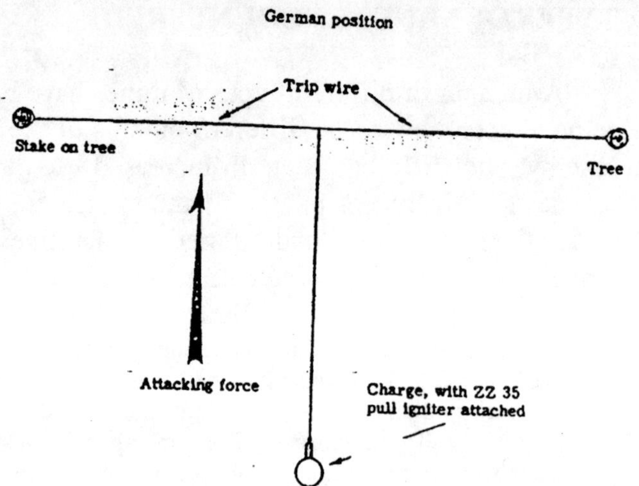

Figure 21.

c. Booby-trapping Telephone Wire

Recently a German patrol operating at night came across an artillery observation post telephone wire, cut it, and booby-trapped the loose ends of the wire in the following manner. The enemy buried two S-mines about 10 yards apart, so that in each case about an inch of detonator showed above the ground. Each of the loose ends of the telephone wire was attached to one of the S-mines with a 12-inch length of fine strong thread. It was the German intention that a linesman would carelessly pick up what he assumed would be a loose end of wire, and thus detonate an S-mine.

At the place where this booby-trapping was done, the telephone line ran quite close to an unimproved road. Four Tellermines were laid in the road, evidently for the purpose of destroying any maintenance truck which might be brought up. The Tellermines were not actually buried, but were covered with mud. The Germans probably realized that neither part of this ruse would offer much danger in daylight, but it was hoped that it would prove effective during the night if prompt maintenance of the line were attempted.

d. Laying Tellermines

As the enemy has been forced back in Italy, he has seized every possible opportunity to mine the most suitable artillery position areas in the terrain he has abandoned. Numerous Tellermines, wooden box mines, antipersonnel mines, and various types of booby traps have been founded in these positions. In one position no fewer than 32 Tellermines were removed from the ground on which a single gun was subsequently emplaced. The commander of a howitzer group recently reported, "In one battery position, a Tellermine was located by a detector. When the usual precautions were taken to disarm and remove the mine, a wire was found which led to 18 cases of dynamite buried in the position area. Each case contained 25 kilograms of explosive."

In many instances the terrain has definitely canalized tank movement, and has enabled the Germans to mine specific routes with every expectation that these would be traversed by tanks and other vehicles.

In one area the Germans had prepared a field of about 200 Tellermines, laying them in a variety of ways. There were no booby traps in this field, but 90 percent of the mines were double Tellermines (one mine laid on top of another), 4 percent were triple, and the remainder single. In all instances of double and triple laying, the mines were placed in close contact with each other, and only the top mine was primed. The purpose of this laying apparently was to increase the size and effect of the charge, which the Germans hoped would destroy entire tanks instead of tracks and bogie wheels alone.

In the case of another field of Tellermines, the Germans went in for triple laying, burying the mines at depths of 6 inches, 2 feet, and 4 feet, respectively. This method evidently was adopted with the expectation that, after the top mine had been located and removed, the others would not be detected.

The Germans have also laid double Tellermines in dirt roads and trains, working from the outside edge of a bank, as illustrated in figure 22a. The enemy intention is that, after a number of tanks have used the road, the slope of the bank will be squeezed over and a rut eventually will be worn to the level of the top of the mine (see figure 22b.)

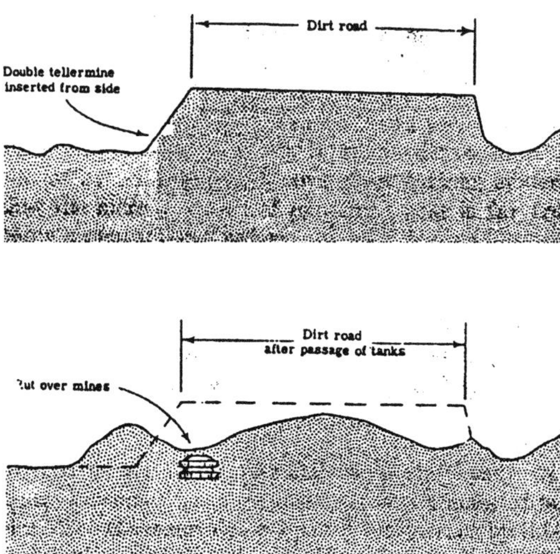

Figure 22. – German Method of Laying Tellermines in
Dirt Roads and Trails

e. Laying S-Mines

S-mines have been found in all kinds of locations, and in quantities varying from a few scattered mines to extensive fields containing as many as 300 mines. One of the many German practices in this respect is the laying of long, narrow anti-personnel mine belts in front of defensive positions. Although mine belts vary greatly in pattern, the following example is a good specimen of enemy procedure. The belt in question consisted of 50 pusher-igniter "S" mines and 50 pull-igniter "S" mines. The mines were laid in four rows, with the rows 5½ yards apart and with the mines in each row 22 yards apart. The whole belt was bordered on each side by a fence consisting of a single strand supported by pickets 18 inches high.

Figure 23 illustrates the layout of the belt.

The stakes for pull-igniters were very noticeable, as were the trip wires. It is interesting to note that the "S" mines set with push-igniters were in outer rows, and "covered" the trip-wires stakes. Many of the push-igniter mines were completely buried.

Recently a detachment of Engineers lifted many S-mines, which had been placed in tracks left in the earth by automobile tires. Apparently a vehicle had been driven over the place chosen for the S-mines, and, once the mines had been laid, they were camouflaged by means of a carefully simulated tire impression which restored the pattern of the original track. The important point to note about this German ruse is that it should dispel any erroneous idea that a vehicle track is sure to be safe. The enemy technique discussed at the end of subparagraph **d** above gives further evidence that the mere presence of existing tracks is far from being a guarantee of safety.

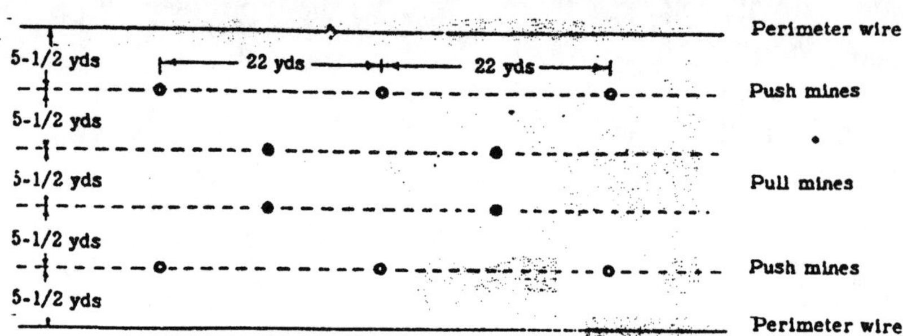

Figure 23. – A German Antipersonnel Mine Belt

In Sicily U.S. soldiers encountered large quantities of S-mines, often very cleverly hidden. A U.S. infantry officer, discussing this subject, said, "The Germans would cover the three prongs of the detonators with loose twigs and grass, so that you couldn't see them at all. At times they laid S-mines by the hundreds in dry stream beds, and covered the prongs with small pebbles, so that it was impossible to know they were there, inasmuch as the whole area was a mass of pebbles. The Germans also prepared booby traps by interlacing the trip wires in the branches and suckers of blackberry bushes. The average soldier would not have suspected the presence of wires at all.

RECENT MINEFIELD DOCTRINE

There is nothing casual or slipshod about the German approach to the business of laying minefields. Each stage of the work is performed with characteristic thoroughness, and complete written records of various types are drawn up for every minefield that is laid. An exception is made only in the case of a minefield laid hastily on the surface of the ground in an area where constant supervision can be maintained, and when early removal is intended. There are four types of German mine records: plans, sketches, reports, and maps. These documents can be extremely useful to Allied intelligence officers, and should be dispatched to them as quickly as possible.

1. Mine plans are descriptions and scale drawings of mine obstacles, containing all necessary details and locations.

2. Mine sketches are simplified mine plans not drawn to scale. These sketches are prepared while a field is being laid, or, if hostile fire or other adverse field conditions interfere, shortly after it has been completed.

3. Mine reports concern scattered mines in an area designated by an Army High Command. Such documents contain information regarding the general location, number, and types of mines used, as well as all necessary data about fuzes. However, no survey work is included.

4. Mine maps show the over-all picture of minefields laid in various sectors of the front, but do not contain detailed information. The maps are based on mine plans and mine reports, and are prepared by the Division Engineer.

German engineers are charged with the preparation of all minefield documents, and with keeping the troops fully informed regarding the mine situation. Mine documents are classified, of course, and it is forbidden to take them into the front lines or to give them to patrols. The German policy of supplying copies of mine documents to higher echelons makes it possible for originals to be destroyed readily if they are in danger of being captured. If a unit has prepared documents is relieved, it turns over the papers to its successor.

MINEFIELD MARKING

German use of minefields to cover defensive actions and withdrawals has been increasing steadily ever since the enemy went over to the defense in North Africa. In a static situation every mine field is regarded as an element of the frontline position, and is surveyed as such. When extensive minefields are prepared, they are laid according to an over-all plan which takes into consideration all weapons and their fields of fire. Mine reconnaissance and the marking and surveying of minefields still are performed exclusively by the engineers.

Mine obstacles can consist of either patterned minefields or of scattered mines, as shown in the illustration. However, it is now an enemy policy that permission must be obtained only from a very high echelon before mines can be laid at random and without any pattern at all.

Extensive fields of pressure-type mines may be subdivided into individual fields, sufficiently far apart to prevent sympathetic detonation. For example, a field of buried Tellermines may be staggered in this manner, with an interval of at least 60 feet between any two of the constituent fields.

Constituent Fields Spaced so as to Prevent Sympathetic Detonation

German systems of minefield marking naturally are intended to enable friendly troops to identify the fields easily, and, at the same time, to keep a hostile force from discovering that mines are present in an area. Stakes, signs, or fences may be used – or all three – depending on the tactical situation and the terrain. Markings are checked and maintained or changed if they have become overgrown or if snow falls, thaws, or heavy rains have made them at all indistinct. As a safeguard against betraying the shape and size of the fields, the Germans stipulate that fences must not follow the outlines of mine fields too exactly. Mine gaps sometimes are marked, especially for patrols and as one of the preliminaries to an assault. For permanent patrols, new gaps are made from time to time, and the old ones are closed. The Germans consider it expedient to start the marking of mine gaps at some distance from each edge of the field. Markings for patrols are maintained for a brief period only; those made as a preliminary to an assault are not prepared until just before the hour when the assault is to take place. All troops are required to be familiar with the various types of survey points, signs, and fences, and are ordered not only to avoid damaging them, but to report when damage of any sort has been observed.

When German engineers prepare a minefield, they mark each corner with a post. After the minelaying has been completed, these posts are driven into the ground until they are almost level with the surface, and then are camouflaged. A mine-free safety strip is provided on the German side of the field.

In addition to trees and other natural reference points for minefield surveying, artificial fix points may be used. Wooden marking stakes are likely to be employed for this purpose. Sometimes stakes are tarred, and their heads reinforced with a metal band, to increase durability. The Germans recommend that all marking stakes for a division or corps area may be made beforehand, to ensure uniformity; it is pointed out that friendly troops will more readily recognize uniform marking stakes, will not remove them, and will find them helpful in crossing mined areas. The length of these stakes varies according to the terrain. The German Army orders that they be flattened on one side for a length of about 8 inches, and that the flat surface be painted red, with the letters *"Mi"* (denoting *"Minen"*) in black. Additional data may be painted on the red surface, if necessary. It is explained that stakes marked in this manner should not be used on the hostile edges of minefields, lest they aid Allied troops in detecting the field.

Mine Stakes (red and black).

Signs (red and white) for Minefields.

Minefield Markers (black and white).

Signs for minefields are painted red and white on boards or sheets of metal, about 1 foot high and 2 feet long, which are then fastened to two stakes. The edges of minefields are marked by signs with horizontal stripes. Edges of gaps through minefields are marked by signs divided vertically. The white half is on the side of the gap; the red half, indicating danger, is on the minefield side.

Signs of this size are used behind the main battle positions. In front of the main positions, smaller signs are used, and the unlettered reverse sides, facing the Allied troops, are planted olive drab.

If red paint is not available for minefield signs, black-and-white signs of the following types may be used instead.

"*Minen*" of course means "mines; "*gasse*" or "*gassen*" indicates mine gaps; "*entmint*" indicates an area which has been cleared of mines.

True Minefield—Dummy Minefield.

Signs indicating true minefields are marked with vertical lettering, but signs indicating dummy minefields may be marked with slanting letters. The Germans permit this distinction to be made known to engineer troops only, on the principal that other troops may betray the nature of the dummy minefields by moving across them.

When marking a gap through a minefield, the Germans mark a path extending some distance into the clear on each side of the field, as shown in the illustration below. This is done to permit assaulting German troops to recognize the exact location of the minefield in ample

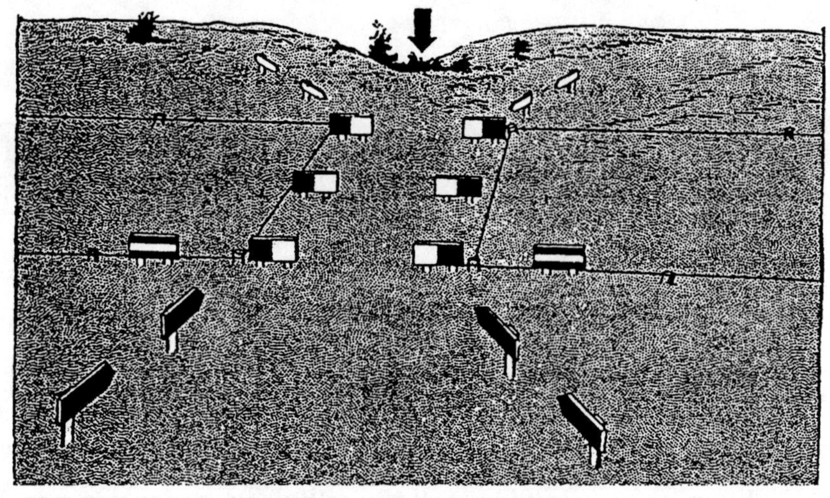

time as they advance, and thus avoid lateral movement in the immediate vicinity of the field.

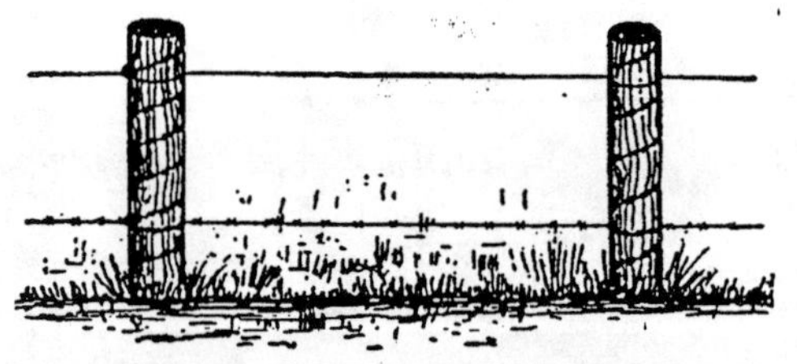

Special Warning Fence.

Minefields are surrounded by warning fences or wire obstacles. (The latter are very often employed in front of centers of resistance.) In the case of an ordinary warning fence, two strands of wire usually are strung between a series of fence posts; the upper strand is of plain wire, and the lower, of barbed wire. Also, barbed wire is wound around fence posts. The Germans try to make all warning fences uniform in a given sector, so that their troops will identify them readily.

USEFUL NOTES ON SURVEYING

Although in general the German Army's methods of surveying minefields are not of particular interest to U.S. junior officers and enlisted men, it should be recognized at times that the enemy may be interrupted in the midst of performing this work and may be forced to abandon an area hastily, leaving behind certain evidence which may prove useful.

It is the enemy's policy to survey all minefields, and the individual mines within the fields, while the minelaying is being done. While this work is in progress, trip-wire obstacles often are marked with small flags, as illustrated. These flags are not removed until all the mines in the field have been laid and are armed.

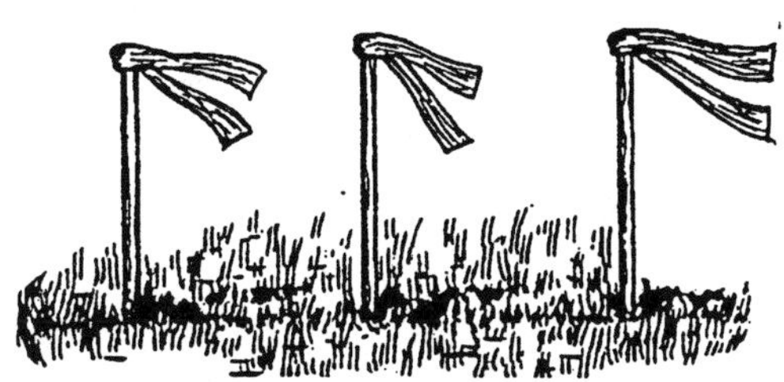

Mine Flags.

All reference points for minefield survey are chosen so that they can be found easily. In some instances the Germans find it necessary to use guide wires.

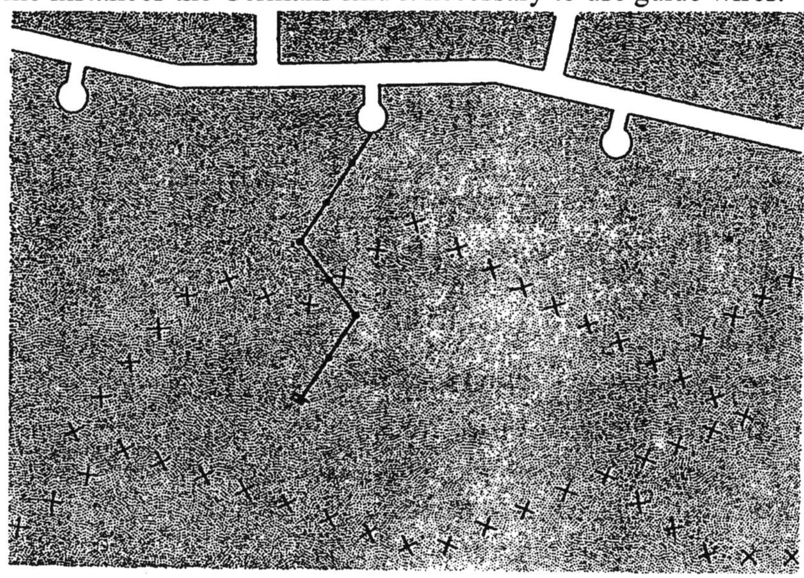

Minefield with Guide Wire.

The Germans try to chose simple fixed points wherever possible – a grade crossing, for example, or the intersection of two roads at the end of a village.

The accuracy of the minefield survey depends on the time available and on the tactical situation. Minefields in front of, or in the immediate vicinity of, a main defense area are surveyed with complete accuracy, to safeguard friendly troops. If a German withdrawal is imminent, a higher command echelon may rule that rough sketches will be sufficient.

When minefields are prepared in the immediate vicinity of a main defense area, the Germans are likely to use trenches, road forks, or pillboxes as reference points. Survey points are established near these reference points.

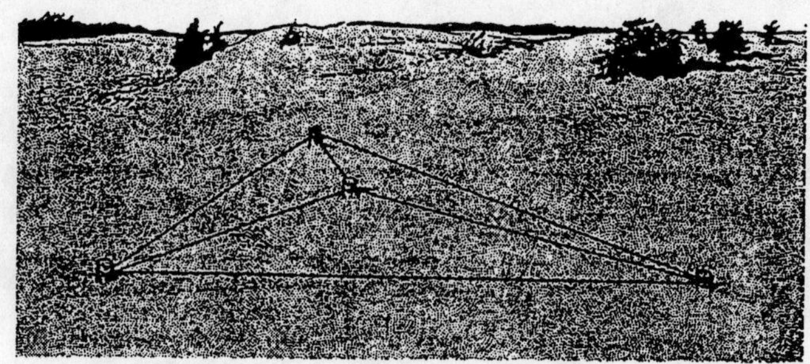

Artificial Survey Point.

Surveying always proceeds from the German side toward the Allied side. Here is a type of survey point *(VP)* that the Germans have found satisfactory. The fix point itself is established at the center of an equilateral triangle with sides 15 to 25 feet long. The corner points and the fix point are stakes, rails, or concrete posts about 3 feet long, which are connected with barbed wire after they have been driven into the ground.

The Germans contend that, since even heavy shelling will hardly ever destroy more than one or two stakes, such a survey point can be reestablished easily.

Guides to Corner, Survey, and Fix Points.

Signs are used to make it easier for troops to locate corner, survey, and fix points. The Germans often consider it advisable to mark certain points by posts point toward them in line of sight. These posts are placed 60 feet apart, and are prepared with one, two, or more grooves—the number increasing with the distance from the minefield. Thus, even if a stake is destroyed German troops still can find the minefield without too much difficulty.

Mine fields arranged in echelons are surveyed by using corner posts on the Allied side of intermediate minefields as survey points.

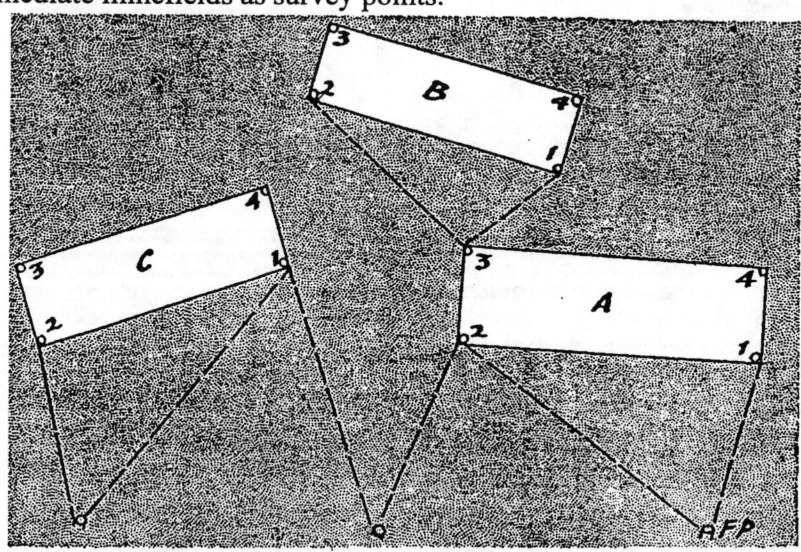

Corner Posts for Minefield Surveying.

Conventional Signs For Mine Maps Forwarded to High Echelons

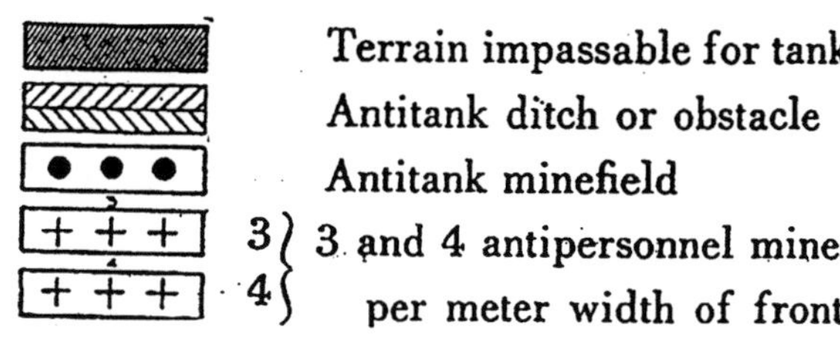

Terrain impassable for tanks

Antitank ditch or obstacle

Antitank minefield

3 } 3 and 4 antipersonnel mines

4 } per meter width of front

Conventional Signs For Mine Plans and Sketches Used By Troops

Tellermines (index numbers used only if different types of Tellermines are laid in the same field)

Topfmines

Riegelmines

Improvised antitank mines

S-mines

Concrete stake mines

Schümines 42

Improvised antipersonnel mines

Small hidden charges

Large hidden charges

Observation mines

Booby-trapped mines

Mines lifted or destroyed

Scattered mines

Deliberate minefield

Mines lying on the surface

Mines below the surface

Minefield cleared or destroyed

Gaps through minefields

Dummy minefields

Built-in hidden charges in buildings

Survey point (*VP*) and fix points (*FP*)

Warning fences

Direction of Allied attack

A GLOSSARY OF GERMAN MINES AND FUZES

Behelf	*Beh.*	Improvised mine
Blind	—	Unarmed
Bohrpatrone	*Bhr. Ptr.*	Demolition cartridge
Brennzünder	*Bz.*	Friction fuze
Brettstückmine	—-	Plank mine
Chemischer Zünder	*Chem. Z.*	Chemical fuze
Doppelzünder	*Dopp. Z.*	Double fuze
Druck	—	Pressure
Druckbrettmine	—	Tread mine
Druckzünder	*D.Z.*	Push fuze
Eismine	*Eis. or Es.*	Ice mine
Entlastungzünder	*E.Z.*	Pressure release
Flasche	*Fl.*	Bottle
Fusschlingemine	—	Snare mine
Geballte Ladung	—	Compact charge
Geschossmine	—	Shell mine
Glas	—	Glass
Glühzünder	—	Electric detonator
Hebel	—	Lever
Holzmine	*H.*	Wooden mine
Kasten	—	Box
Kipp	*Ki.*	Tilt
Knallzündschnur	—	Detonator fuze
Knick	—	Snap
Ladung	—	Charge
Leicht	*l.*	Light
Mine	*Mi.*	Mine
Panzermine	*Pz.*	Antitank Mine
Papp	—	Cardboard
Pilz	—	Mushroom
Riegel	*R.*	Bar
Schlag	—	Impact
Schupo		Policeman
Schümine		Antipersonnel mine
Schützen	—	Infantry
Schützenmine	*S.*	Antipersonnel mine
Scharf	—	Armed
Schnell		Quick
Schwer		Heavy
Sicher		Safe
Sprengbüchse		Slab charge
Sprengkapsel		Detonator
Sprengkörper		Small charge
Sprungmine		Jumping mine
Stab		Rod
Stock		Picket, stake
Stolperdrahtmine		Trip-wire mine
Stück		Piece, part
Teller	*T.*	Plate
Topf	*T.*	Saucepan
Verzögerung	—	Delay
Zeit	*Zt.*	Time
Zeitzündschnur		Safety fuse
Zug	*Z.*	Pull
Zugdruckzünder	*Z.D.Z.*	Push-pull fuze
Zug-und Zerschneide-zünder	*Z. u Z.Z.*	Pull and tension release fuze
Zugzünder	*Z.Z.*	Pull fuze
Zünder	*Z.*	Fuze[1]
Zündschnuranzünder	*Zdschn. Anz.*	Safety fuse lighter
Zwischenstück		Adaptor

[1] In British terminology: "igniter.

INDEX

Other Works Available From
The Nafziger Collection, Inc.
PO Box 1522
West Chester, OH 45069-1522
http://home.fuse.net/nafziger

PRICING: All works are softbound. Unless otherwise noted in all of the following works, the price of individual volumes is $19.95, plus $2.50 S&H for 1st book, $2.00 for a second, and $1 for each book thereafter. Shipping costs are for continental US only.

WORLD WAR II TACTICAL MANUALS The five US studies are reprints of various World War II field manuals. They provide everything you would want to know about US armored tactics: bivouacs, offensive actions, defensive actions, use as artillery, formation diagrams, numerous illustrations, weapons characteristics, and much more. They are ideal for the individual who wants a better understanding of US tactics from the company to the battalion level, be they a historian or a wargamer.

AMERICAN TANK COMPANY TACTICS (FM 17-32): A reprint of the 1944 original government publication. One volume.

AMERICAN ARMORED BATTALION TACTICS (FM 17-33): A reprint of the 1944 original government publication. One volume.

EMPLOYMENT OF TANKS WITH INFANTRY (FM 17-36): A reprint of the 1944 original government publication. One volume.

US ARMORED INFANTRY BATTALION TACTICS IN WWII (FM 17-42): A reprint of the 1944 original government publication. One volume.

EARLY ORGANIZATION AND TACTICS OF AMERICAN TANK DESTROYER UNITS (FM 17-5): A reprint of the 1942 original government publication. One volume.

AMERICAN RECONNAISSANCE TACTICS IN WWII (FM 17-20 & 22): A reprint of the 1944 original field manuals on battalion, company and platoon reconnaissance techniques.

AMERICAN ARMORED DIVISION TACTICS IN WWII (FM 17-100): A reprint of the 1944 original and the 1943 proposed manuals on divisional operations.

TACTICS OF THE NORMAN HEDGEROWS (JUNE-AUGUST 1944): By G.F.Nafziger, 83 pages.
     This work focuses on the tactical experience in Normandy as the US fought its way through the Norman hedgerows and the Germans fought to stop them. It is based on American intelligence and other reports of the tactics that worked and the German defensive tactics. It shows an evolution of tactics in addition to a variety of tactics employed by specific units, all of which developed their own systems.

GERMAN TACTICS IN WESTERN EUROPE 1943-45: By G.F.Nafziger.
     Same as German Tactics in North Africa, except the reports cover mostly Western and some Eastern Europe tactics. Heavily illustrated.

GERMAN TACTICS IN NORTH AFRICA: By G.F.Nafziger,

This is a compilation of a number of US intelligence documents that have been screened and organized. The documents are provided with the original citations, allowing them to be traced to those original documents. Some intelligence shortcomings are corrected, but otherwise the document is faithfully reproduced. Heavily illustrated.

COMPARATIVE PERFORMANCE OF GERMAN AND ALLIED ANTI-TANK WEAPONS IN WWII: Edited by G.F.Nafziger, 47 pgs, $12.95 plus S&H.

This is a statistical examination of the causes of tank losses and a comparison of the effectiveness of German versus American and British anti-tank weaponry.

THE GERMAN TANK PLATOON IN WWII: ITS TRAINING AND EMPLOYMENT IN BATTLE: Edited By G.F.Nafziger. 114 pgs, 179 illustrations.

This is a reproduction of the 1941 US Army translation of the 1940 German tactical training manual. It contains a forward By General Heinz Guderian and was prepared incorporating the lessons learned By the Germans in the 1939 Polish and 1940 French Campaigns. It is heavily illustrated with pictures and diagrams of the right and wrong way to operate. It was designed as the basic training manual for the German panzer crew. A highly informative work for either the historian or the gamer. Comments By Gulf War veteran tankers say that the lessons it teaches are still taught today.

GERMAN PANZER TACTICS IN WORLD WAR II, Combat Tactics of German Armored Units from Section to Regiment: By Charles C. Sharp

This work contains 101 pages chocked full of information. It provides detail son the fundamentals of German tactical doctrine, training, detailed organizations of panzer units for each campaign, and then begins a series of "tactical lessons" drawn from various campaigns, beginning with Poland and examining France, Africa, the Russian front in 1941, the operations during Kursk and facing the British and Americans in 1944.   There are numerous illustrations of tactical panzer formations and discussion on how they were used.

GERMAN SQUAD TACTICS WORLD WAR II,   By Matthew Gajkowski. 106 pages,

The first portion of this unusual work is a detailed review of German squad level infantry tactics using German regulations and American studies. It is highly illustrated, and provides tactical exercises to illustrate points.   The second portion is a translation of the German Panzer Grenadier tactics and reviews both mounted and dismounted tactics. Diagrams show the proper deployment both of dismounted infantry and half tracks in combat. The last portion of the panzer grenadier section even provides illustrations and explanations of hand signals used for communicating between half tracks. The final portion of this work is a series of highly detailed TO&E listings of various German infantry companies from 1943 through 1945, including panzer grenadier companies, grenadier companies, bicycle companies and skijäger companies.

SOVIET TACTICAL DOCTRINE IN WWII, 146 pgs.

This is a reprint Chapter V of the extremely rare US "Handbook on U.S.S.R. Military Forces TM-30-340. It contains numerous diagrams of regimental, corps and army level tactics, supported By dozens of illustrations. The work was prepared By the US War Department using all the information that it could obtain from General Gehlen and his staff of Fremde Herre Ost ("Foreign

Armies East").   It is a comprehensive examination of Soviet ground tactics, including the important support elements.   It covers Soviet tactical command structure, the role of combat intelligence and reconnaissance, the tactical planning process and documentation, provides an examination of tactics proper, including infantry tactics, a huge and highly detailed examination of artillery support, armored and cavalry tactics, and engineering and anti-aircraft support, special operations, including night operations, the use of smoke assaults on fortified or urban areas, river crossings, actions in woods, swamps and mountains, and winter warfare.   The emphasis in all sections is on the last part of the war, but much of the material applies throughout the war.

SOVIET ARMORED TACTICS IN WWII, THE TACTICS OF THE ARMORED UNITS OF THE RED ARMY FROM INDIVIDUAL VEHICLES TO BATTALIONS ACCORDING TO THE COMBAT REGULATIONS OF FEBRUARY 1944:   By Charles C. Sharp.

SOVIET INFANTRY TACTICS IN WORLD WAR II; :Red Army Infantry Tactics from Squad to Rifle Company from the Combat Regulations   By Charles C. Sharp,

WORLD WAR II CAMPAIGN AND OTHER STUDIES

THE WESTWALL (SIEGFRIED LINE) 1938-1945,   By Jean-Denis Lepage, 94 pages.

This work is very heavily illustrated, with numerous B&W drawings of exteriors and interiors of every different standard bunker and pillbox used By the Germans in the Siegfried Line. After a brief history of fortifications, This work explores the history, the construction, and the elements of the Siegfried Line. If you've any interest in building authentic models for your gaming table or just seeing the variety and uses of the bunkers and other works that formed the Siegfried line, this book is for you.

THE AFRIKA KORPS, AN ORGANIZATIONAL HISTORY 1941-1943,
By G.F.Nafziger.

This work is organized into four parts. The first is the history of the organizational evolution of every German division that fought in Africa. The second part is a collection of orders of battle drawn from OKH and Afrika Korps records listing all units, including corps combat, support and supply elements. The third section contains the OKH theoretical organization for the Afrika Korps, paralleling the second part. The fourth part is a listing of daily tank inventory and equipment returns for various periods during the north African campaign.

LESSONS LEARNED BY THE AMERICAN ARMY DURING THE TUNISIAN CAMPAIGN: reprint of 1943 original government publication. 32 pages. Single volume for $12.95 plus $2.50 S&H.

This work explores the problems encountered By the American Army during the Tunisian campaign, cites specific examples and quotes individuals involved. Very insightful into the tactical problems that the American army had and how it went about correcting them.

THE SOVIET ORDER OF BATTLE SERIES: By Craig Crofoot.

These four volumes are drawn from the official Soviet war records and provide complete orders of battle for the field armies of the Soviet Union. All formations, including independent Army and Corps units are listed.

The Berlin Direction: April-May 1945, An Extraction of the Official Soviet Army Order of Battle of the Berlin Strategic Offensive Operation April-May 1945 - 1 volume, 104 pages

The Soviet Order of Battle: Volume 1, Part 1: The Northern Theater of Operations, 22 June 1941 - 1 July 1942, 123 pages
The Soviet Order of Battle: Volume 1, Part 2: The Northern Theater of Operations, 1 July 1942 - 1 April 1943, 118 pages

The Soviet Order of Battle: The Sleeping Bear, Vol 1: 22 June 1941. 127 pages. Highly detailed order of battle with complete inventory of aircraft By types, tanks By types, artillery By types, etc., etc.

WORLD WAR II ARMY ORGANIZATIONAL STUDIES

The following works are highly detailed studies of various European WWII armies. These books were developed using archival and primary sources never before available to the English reading public. Each volume provides a history of varying length, a paragraph to several pages, on each unit under consideration. Organizational data, including the numbers and types of weapons, etc., are provided. They will, most assuredly, answer every question you ever had about these armies. These works are soft bound and average about 100 pages per volume.

SOVIET ORDER OF BATTLE WWII: By Charles C. Sharp. Individual volumes $19.95.

Vol 1 - "The Deadly Beginning", Soviet Tank, Mechanized, Motorized
        Divisions of 1940-1942
Vol 2 - "School of Battle" Soviet Tank Corps and Tank Brigades,
        January 1942-1945
Vol 3 - "Red Storm", Soviet Mechanized Corps and Brigades and Guards
        Armored Units, 1942-1945
Vol 4 - "Red Guards", Soviet Guards Rifle and Parachute Infantry
        Units, 1941-45
Vol 5 - "Red Sabers", Soviet Cavalry Corps, Divisions and Independent
        Brigades, 1941-45
Vol 6 - "Red Thunder", Soviet Artillery Corps, Divisions, and Brigades
        Including Rocket, Anti-tank, and Mortar Units, 1941-45
Vol 7 - "Red Death", Soviet Mountain, Naval, NKVD, and Allied Divisions
        and Brigades, 1941-45
Vol 8 - Soviet Rifle Divisions formed up to 22 June 1941
Vol 9 - Soviet Rifle Divisions formed June - December 1941
Vol 10- Soviet Rifle Divisions formed 1942 to 1945
Vol 11- "Red Volunteers", Soviet Rifle and Ski Brigades and Militia
        Units, 1941-45
Vol 12- "Red Hammer", Soviet Self Propelled Artillery and Units
        Equipped with Lend-lease Armor.

THE BRITISH ARMIES OF THE SECOND WORLD WAR - AN ORGANIZATIONAL HISTORY:

  By David Hughes, James Groshot & Alan Philson
Vol 1 - Armoured and Cavalry Divisions - 123 pages
Vol 2 - Dominion Armored Divisions and British Infantry Divisions - 132 pages
Vol 3 - Territorial Army, War Raised Mountain and Training Infantry Divisions
        - 122 pages.
Vol 4 - Tank & Armoured Brigades, 79th Armoured Division, Armd Car Regiments,
        African, Malayan & other Colonial Forces - 126 pages
Vol 5 - The Australian Army - 108 pages
Vol 6 - The Canadian Army - 106 pages

Vol 7 - New Zealand & South Africay Armies - 122 pages
Sup 1 - OB for the British Army in the early part of the war.
Sup 2 - OB for the British Army 1941-1942 - 109 pages
Sup 3 - OB for the British Army 1942-1944 - 97 pages

FRENCH ORDER OF BATTLE IN WORLD WAR II, By G.F.Nafziger

RUMANIAN ORDER OF BATTLE IN WORLD WAR II, By G.F.Nafziger

THE AMERICAN ARMY IN WORLD WAR II: By George F. Nafziger. Individual volumes, $19.95 each.

Vol 1 - 1st-40th Infantry Divisions (includes history of building of army
        from 1939-1945, plus tables of organization and organizational
        history of 1st-40th Divisions.)
Vol 2 - 41st-104th Infantry and Airborne Divisions (only organizational
        history of individual divisions)
Vol 3 - Independent Armored, Mechanized Cavalry, Artillery, Anti-aircraft
        and Engineer Battalions
Vol 4 - Armored Divisions

JAPANESE ORDER OF BATTLE WWII: By John L. Underwood, Jr. Individual volumes
$19.95.
Vol 1 - Infantry & Depot Divisions
Vol 2 - Airborne, Raiding, Amphibious, Cavalry and Independent Brigades
Vol 3 - Independent Battalions, Army Detachments, and Miscellany
Vol 4 - Japanese Armored Units in WWII - 85 pages
Vol 5 - Japanese Armor in Manchuria - 108 pages

JAPANESE INFANTRY ORGANIZATION & TACTICS IN WWII : Edited By G.F.Nafziger.
$19.95 per volume

        The basis of this work is a series of monographs published by the US
Army in the mid-1950's. Its authors are senior Japanese officers with particular insight into Japanese tactics. This is an incredibly insightful work and
is heavily illustrated with all sorts of examples of tactics and exercises.

Vol 1 - Basic Infantry Organization & Discussion of Overall Tactical
        System, including Training Cycle and Night Vision Techniques.
        - 135 pages
Vol 2 - Extracts from Japanese Tactical Manuals & US Intelligence Observations
        on Japanese Defensive Cave Tactics - 138 pages
Vol 3 - Historical Examples of Japanese Night Attacks - 103 pages

ITALIAN ORDER OF BATTLE IN WORLD WAR II, By G.F.Nafziger. Individual
volumes $19.95.
Vol 1 - Armored, Motorized, Airborne & Alpini Divisions
Vol 2 - Infantry Divisions
Vol 3 - Black Shirt, Mountain, & Assault Divisions and the 1944 Italian
        Army in Allied Service

**THE CHINESE ARMY, THE GROWTH AND ORGANIZATION OF THE CHINESE ARMY FROM 1895
TO 1945**: By George F. Nafziger

BULGARIAN ORDER OF BATTLE IN WORLD WAR II: By G.F.Nafziger

SOVIET INFANTRY TACTICS IN WORLD WAR II: By Charles C. Sharp

Developed using Soviet tactical regulations, this 122 page work examines the tactical employment of the Soviet rifle squad, platoon and company, plus the submachine gun squad, machine gun sections, anti-tank rifle squad, and infantry guns. It contains appendices discussing infantry weapons, signals, target maps, a sample reconnaissance diary, anti-tank defenses, sapper work, and tactical symbols. Range and other details are provided for infantry weapons & there are TO&Es for infantry squads, platoons, and companies.

NOTES ON INFANTRY TACTICS IN KOREA By S.L.A.Marshall and appendix by B.H.Liddell Hart. 102 pages.

A second formerly secret study By SLA Marshall on infantry tactics in Korea. This particular work focuses heavily on Chinese tactics and techniques used By the US to counter them. There is an exceptional debriefing of a US infantry company after it executed a bayonet attack against the Chinese that is used to explain the effectiveness of the bayonet when used against the Chinese.

COMMENTARY ON INFANTRY OPERATIONS AND WEAPONS USAGE IN KOREA (WINTER OF 1950-51) By S.L.A.Marshall. 124 pages.
Declassified in 1998, this US Army document discusses General Marshall's observations of weapons usage and effectiveness during the early years of the Korean War.

OTHER WORLD WAR II

ROMMEL AND GUDERIAN AGAINST THE BELGIAN CHASSEURS ARDENNAIS, The Combats at Chabrehez and Bodange, 10 May 1940, Translated By G.F.Nafziger, 123 pages, 3 maps.

This is a translation of a post-WWII Belgian staff study on the heroic stand of the Belgian bicycle troops in the Ardennes as they faced the armored might of Rommel and Guderian. Read how a single Belgian company stopped Rommel's advance cold for six hours. Contains orders of battle.

TANK DESTROYERS IN COMBAT, Edited by G.F.Nafziger, 49 pgs, $12.95.

This is two US Army works published towards the end of WWII that provide details on specific actions of tank destroyers in combat. Though the bulk of the work is focused on the Tunisian campaign, it contains short accounts of operations against the Japanese and in Italy. It also contains a section of lessons learned.

NAPOLEONIC TRANSLATIONS AND OTHER UNUSUAL WORKS By G.F.Nafziger

RUSSO-TURKISH WAR OF 1806-1812, Translated By A.Mikaberidze, $25.00 per volume plus S&H.

Translated from the Russian original By Lt. General Alexander Mikhailovsky-Danilevsky, this work is the ONLY work in the English language on this obscure, yet important war. Volume 1 covers the period 1806 through the end of 1809. It contains 149 pages and 1 map of the theater and 15 maps of the various land and naval battles. This volume goes into great detail on the various battles, providing orders of battle for the Russian Army at the various battles and highly detailed maps to support the text. It also includes a discussion of several naval battles and landings in the Aegean.

Volume 2 picks up with the beginning of the 1810 campaign and continues in the format of the first. The text is supported By numerous maps and orders of battle for the Russians. This particular volume contains numerous illustrations of Turkish and Russian uniforms.

A CONTEMPORARY ACCOUNT OF THE 1796 CAMPAIGN IN GERMANY AND ITALY, Edited & annotated By G.F.Nafziger, 141 pages.

The basic work is *The History of the Campaign of 1796 in Germany and Italy* By Lieutenant General Thomas Graham, who served as a liaison officer with the Austrian army during 1796 in Italy. This is the first of five volumes By Graham. It provides a detailed discussion of the French and Austrian operations in Germany and Northern Italy. All battles, sieges, and operations are discussed. It is the first of five volumes on the 1796-1799 campaigns.

A CONTEMPORARY ACCOUNT OF THE 1797 CAMPAIGN IN GERMANY AND ITALY, Edited & annotated By G.F.Nafziger, 117 pages.

A CONTEMPORARY ACCOUNT OF THE 1799 CAMPAIGN IN GERMANY AND SWITZERLAND, Edited & annotated By G.F.Nafziger, 140 pages.

A CONTEMPORARY ACCOUNT OF THE 1799 CAMPAIGN IN ITALY, Edited & annotated By G.F.Nafziger,   125 pages, 3 maps.

A CONTEMPORARY ACCOUNT OF THE 1799 CAMPAIGN IN HOLLAND, Edited & annotated By G.F.Nafziger,   171 pages, 3 maps, $25.00 plus $2.50 S&H,

OPERATIONS OF THE POLISH ARMY DURING THE 1809 CAMPAIGN IN POLAND, Translated By G.F.Nafziger, 175 pages, 1 map. $25.00. plus $2.50 S&H.

Originally published in 1841 and written By Polish General Roman Soltyk, this work describes in great detail Poniatowski's operations in the defense of the Grand Duchy of Warsaw when it was invaded By the Austrian army under Archduke Ferdinand in 1809.

NAPOLEON'S SYSTEM OF WAR, By General H. Camon, Translated & annotated By G.F.Nafziger.

Initially published in 1923, this work By a French general examines Napoleon's system of war, both strategic and tactical, identifies Napoleon's system of war, and examines it in light of several battles. This translation also contains a translation of a small pamphlet By Camon entitled *"Napoleonic Maneuvers of Cavalry."*

MEMOIRS TO SERVE THE HISTORY OF THE 1796 CAMPAIGN IN GERMANY , By Général de division Jourdan, Edited & Annotated By G.F.Nafziger, 126 pages.

Général de division Jourdan commanded the Army of the Sambre and Meuse in the 1796 campaign. This is his account of the campaign, written as a rebuttal to a history of the campaign written By the Archduke Charles, his opponent in that famous campaign. This work contains a number of orders of battle and many original documents that he uses to document his position.

ARCHDUKE CHARLES' 1796 CAMPAIGN IN GERMANY , By Archduke Charles, Translated by G.F. Nafziger, 222 pgs, $25.00

This is the 1796 campaign in Archduke Charles' own words, translated into French and annotated by Jomini, then translated into English. It is the work that provoked Jourdan to write his account of the campaign. The Archduke is surprisingly critical of his own mistakes and makes no bones about those made by Jourdan and Moreau. It provides an excellent perspective of the 1796 campaign from the Austrian side.

<u>**THE EYLAU-FRIEDLAND CAMPAIGN OF 1806-7**</u>, By G.F.Nafziger, 103 pages.

This work begins with a detailed organizational study of the regiments of the Russian and French armies, providing complete organizational and uniform details. It then begins a general review of the course of the campaign, discussing the various battles, supporting those discussions with numerous campaign and battle maps, plus complete orders of battles for both armies at Eylau and Friedland.

<u>**BRUNE'S 1807 CAMPAIGN IN SWEDISH POMERANIA**</u>, translated and compiled By G.F.Nafziger, 73 pages

This work is a translation of *Historical Account of the 1807 Campaign In Swedish Pomerania of the Observation Corps of the Grande Armée commanded by Maréchal Brune followed By a Review of the Maréchal*, By the Chevalier Vigier de Saint-Junien, published in 1825. Attached to it are a number of translations from various sources of the actions By the various non-French contingents, including the Spanish, Dutch, and various German states during the campaign and siege of Stralsund. It is supported By a number of orders of battle from various sources, including the French archives. Two maps are included.

<u>**HISTORICAL ACCOUNT OF THE MILITARY OPERATIONS OF THE ARMY OF ITALY IN 1813 AND 1814**</u>, translated By G.F.Nafziger, 93 pages

Originally By Général de division Comte de Vignolle, the chief of staff to Eugène Beauharnais, the commanding officer of the Army of Italy in 1813 and 1814, this is an account of French operations against the combined actions of Austria, England and Naples. It contains several orders of battle for the French and a detailed map of the battle of Mincio.

<u>**AN ENGLISH GENERAL IN THE ARMY OF REVOLUTIONARY FRANCE 1792**</u>, Annotated By George F. Nafziger, 84 pages.

Originally published in 1794, this is the personal account of General John Money, the ONLY English general to serve in the armies of Revolutionary France. He accepted a commission in the French army in early 1792 and arrived in France in time to participate in the seizure of the Tuileries By the Parisian mob on 10 August 1792. It goes on to discuss Money's military operations under Generals Dumouriez, Dillon and Valence in northern France. The work contains 3 maps.

<u>**THE FIRST PHASE OF NAPOLEON'S 1796 CAMPAIGN**</u>, Translated & annotated By George F. Nafziger, 103 pages, $25.00.

Originally published By the Historical Section of the French General Staff in 1905, this work is a collection of reports written By French officers after field surveys made in 1799. It covers the battles of Montenotte, Dego, Ceva and others. The work focuses on the Piedmontese Army and its operations. It is highly detailed, providing deployments of the Piedmontese army at the various battles, descriptions of the field works prepared for those battles, detailed discussions of the battles and of the terrain in general. The work has had one map added to it for the benefit of the readers.

<u>**WARS OF THE FRENCH REVOLUTION, Volume 1, 28 April 1792-27 August 1793**</u>, Translated By George F. Nafziger, 153 pages, 9 maps, $25.

This is the first in what will be a series of 10-12 volumes covering the wars of the French Revolution. It is a translation of the 1900 edition of a work originally published in 1818 under the title *Victoires, conquêtes, désastres, reverset guerres civiles des Française de 1792 à 1815*, By the Society of

Soldiers and Men of Letters. It is a chronological discussion of the campaigns, maneuvers, and, most particularly, the battles. It covers all the major battles and most of the small engagements. This particular volume discusses battles and skirmishes in the Pyrenees and devotes considerable time to discussing the battles in the Vendée.

<u>WARS OF THE FRENCH REVOLUTION, Volume 2, 28 August 1793-26 May 1794</u>, Translated By George F. Nafziger, 148 pages, 7 maps, $25.
This is a continuation of the previous work.

<u>HISTORICAL DICTIONARY OF THE FRENCH REVOLUTION AND TERROR</u>, By George F. Nafziger. One volume - 103 pages, index & bibliography.
With nearly 600 entries, this work provides short biographies or explanations of the individuals and events of the French Revolution (1789-1799). It lists every general executed, all the major politicians and many of the minor ones, every significant terrorist, the various governmental structures, the significant battles, and other subjects that will provide the reader with a very thorough understanding of the history of the French Revolution and Terror.

<u>A TREATISE UPON THE REGULATIONS OF THE FRENCH INFANTRY</u>, By Général de brigade Meunier. One volume.
This work was originally published in 1805 and an English translation was published in 1809. This is a reprint of the 1809 translation plus biographies of Général Meunier and Chef de brigade Dedon, for reasons that are explained in the book; plus an explanation of the organization of the French infantry and other details that permit a fuller understanding of the French infantry drill system that was proposed By Meunier.

<u>THE FIRST PRUSSIAN INVASION (11 AUGUST-2 SEPTEMBER 1792)</u>, By Arthur Chuquet, translated and annotated By G.F.Nafziger, 177 pages, $25.00.
This is the first in a series of 11 works written By Arthur Chuquet on the early years of the French Revolution. This particular volume is a highly detailed account of the sieges of Longwy and Verdun, and all the events leading up to the battle of Valmy. It is filled with details on every aspect of military operations of the early days of the wars of the French Revolution.

<u>VALMY</u>, By Arthur Chuquet, translated and annotated By W.D.Peterson, 164 pages, $25.00.
This is the second volume in the Chuquet series on the Wars of the French Revolution. It focuses on the Prussian advance to and the battle of Valmy. Not only does it examine the battle in detail, but it also discusses the actions of the surrounding columns that were not at Valmy. It is a the finest work on this aspect of the wars of the Revolution that I've ever seen.

<u>ESSAYS ON THE NAPOLEONIC WARS, Volume 1</u>, by various authors, 80 pages.
This is a collection of essays by experts in the Napoleonic era that were submitted to the Napoleon Forum as part of essay writing contests. They cover a myriad of topics: Russian casualties at Austerlitz, Davout's orders prior to Auerstadt, Bagration in the 1807 campaign, French artillery at Friedland, Walcheren, Uniforms of the Portuguese 6th Cacadores, Combat at Majahalonda 11 August 1812, Russian artillery at Borodino, British Manpower problems in the Napoleonic wars, and the British regimental mess.

NAPOLEONIC STUDIES By G.F.Nafziger

This series of booklets examines the army of each nation, providing a complete history of its organization, the raising, disbanding, and the modification of the organization of every unit in that army. Details are provided on uniforms and their changes, in support of the discussion of organization. In addition, to a varying degree, the battle history of the army and its units is discussed. These booklets were designed to provide the wargame gamer with everything that might be needed to build any of these armies, but they are also the finest, and in most cases, the only history of these armies in the English language. They were prepared using archival materials, official documents ranging from regimental histories to governmental studies, and memoirs of members of these armies. These works are soft bound and <u>average</u> about 100 pages in length.

TACTICS OF THE RUSSIAN ARMY IN THE NAPOLEONIC WARS, By Alexander
& Yurii Zhmodikov, Each volume is $19.95.

Vol I - Russian Army in the 18th Century & Tactical Response to the French
        Revolution, 1801-1809 - 117 pages
Vol 2 - Tactical Changes in Response to Napoleon - 1810-1814 - 125 pages

An exhaustive study, drawn from period Russian regulations, eye-witness accounts, period professional journals, and official reports, as well as later documentation, this two volume work covers absolutely every aspect of the Russian system of war from 1792-1815. It explodes all the myths and misconceptions that have haunted the literature on the Russians and sets the record straight.

ROYAL, REPUBLICAN, IMPERIAL, A HISTORY OF THE FRENCH ARMY FROM 1792-1815, By
George F. Nafziger. Individual volumes $19.95.

Vol 1 - Infantry - History of Line Infantry (1792-1815), Internal
        & Tactical Organization; Revolutionary National Guard,
        Volunteers Fédérés, & Compagnies Franches; and 1805
        National Guard.
Vol 2 - Infantry - National Guard after 1809; Garde de Paris,
        Gendarmerie, Police, & Colonial Regiments; Departmental
        Reserve Companies; and Infantry Uniforms
Vol 3 - Cavalry; Line, National Guard, Irregular, & Coastal Artillery,
        Artillery & Supply Train, and Balloon Companies.
Vol 4 - Imperial Guard
Vol 5 - Foreign Regiments - Swiss, Spanish, King Joseph's Army,
        Croatians, Portuguese, Régiments d'étranger, Albanians,
        Hanoverian Legion, etc.

THE PRUSSIAN ARMY DURING THE NAPOLEONIC WARS (1792-1815),
By George F. Nafziger. Individual volumes $19.95.
Vol I   - Infantry (1792-1815)
Vol II - Guard & Landwehr (1792-1815)
Vol III - Cavalry, Artillery, Technical Troops, & Train (1792-1815)

THE CRESCENT AMONG THE EAGLES OTTOMAN EMPIRE AND THE NAPOLEONIC WARS, By
W.E.Johnson

THE IMPERIAL RUSSIAN ARMY, 1763-1815,
Vol I -   Infantry, Mixed Units, Legions, Marines, Engineers, Pioneers, Sappers
          & Opolochenie
Vol II - Cavalry, Cossacks, Guard, and Artillery

THE BRITISH ARMY, ITS REGIMENTS AND BATTALIONS, 1793-1815,   By Michael G. McKen-
na. $25 per volume

Vol I - Administration & Organization, Civil Departments, Officers, Cavalry,
         Infantry and Fencible Units - 148 pgs
Vol II - Veteran & Garrison Units, Ordnance, Colonial Units, Foreign Units,
         KGL, Royal Marines, Militia & Volunteers, and the Honourable East
         India Company - 142 pgs

**THE ARMIES OF THE KINGDOM OF BAVARIA AND THE GRAND DUCHY OF WÜRZBURG 1792-
1815**

THE ARMIES OF SPAIN AND PORTUGAL, 1808-1814,

THE WÜRTTEMBERG ARMY, 1792-1815,

THE ARMIES OF BRUNSWICK, HANOVER, HESSE-CASSEL, AND THE HANSEATIC CITIES.

THE ARMIES OF WESTPHALIA AND CLEVES-BERG, 1806-1815

ARMIES   OF THE GERMANY AND THE CONFEDERATION OF THE RHINE 1806-1815 Vol I.
(Anhalt, Saxe-Gotha, Saxe-Hildburghausen, Saxe-Meinigen, Coburg, Frankfurt
Primate, Hesse-Darmstadt, Nassau, Mecklenburg-Schwerin and Mecklenburg-Stre-
litz)   -

ARMIES   OF THE GERMANY AND THE CONFEDERATION OF THE RHINE 1792-1815 Vol II.
(Baden, Lippe, Reuss, Schwarzburg, and Waldeck)

THE PRUSSIAN ARMY OF FRIEDRICH DER GROßE, By G.F.Nafziger, $19.95 per vo-
lume.

Volume 1 - The Infantry - 92 pages
Volume 2 - The Saxons, Cavalry & Technical Troops - 92 pages

SEVEN YEARS WAR ARMY STUDIES

THE NEW USE OF ARTILLERY IN FIELD WARS: NECESSARY KNOWLEDGE, By Chevalier
du Teil, translated by Charles Shallcross, 74 pages.

Originally published in 1778, this was one of the early and most influential
essays on the tactical employment of field artillery in warfare immediately
before the French Revolution. It encompasses the lessons and experiences of
the Seven Years Wars and laid the foundations for what would occur in the
Napoleonic era.

THE FRENCH ARMY OF LOUIS XV, In the War of the Austrian Succession and Seven Years War , By
Lucien Mouillard, Translated and Annotated by G.F.Nafziger, $50, 152 pages

This is THE classic work on the French Army during the War of the Austrian
Succession and Seven Years War. It contains all organizational and structural

details of the army as evolved over the 18th century. It also contains 57 color plates that provide details of the uniforms and flags of every regiment and formation. This work is a must for anyone interested in Louis XV's Army.

OTHER ERAS

GERMAN NAVAL INFANTRY IN WORLD WAR I, By Dirk Rottgardt, 141 pages, $25.

This is a painstakingly researched study of German naval infantry in World War I. It includes not only organizational and operational details of these troops, from their inception, but lists all the coastal batteries they manned, as well as the numbers and types of guns in each battery. It explains how each individual naval rank and rating functioned, requirements for promotion, training, etc., etc. Its author is a German who had access to numerous archival and other documents with which to prepare this work.

The Nafziger Collection
Orders of Battle Through the Centuries
PO Box 1522
West Chester, OH 45069-1522

The Nafziger Collection is a constantly growing mass of orders of battle covering the period from 1600 to 1945. It started because of a personal interest in the Napoleonic Wars, but steadily grew into other areas because of the gaming public's interest in these highly detailed, accurate historical orders of battle. Sources range from published works to actual archival documents, which represent the largest single source. Nearly all OB's break down to the regimental level. The availability of strength figures and artillery equipment varies from period to period.

The catalog comes in sections for your convenience. All of the catalog sections are available for $6.00. The cost of the individual sections are listed below. Prices of the OB's are $0.25 per page, with discounts on large orders and multi-paged OB's. Full details on ordering come with the catalogs. If you have access to the INTERNET you can download the catalogs for free from my home page "http://home.fuse.net/nafziger" and always get the latest.

<u>17th Century Catalog</u>
> 1600-1699, including
>> 30 Years War & English Civil War
>> Total 150 OB's
>> Catalog Price $0.50 - 2 pages

<u>18th Century Catalog</u>:
> 1697-1792
>> Spanish Succession Great Northern War
>> Polish Succession, Austrian Succession
>> Seven Years War, Bavarian Succession
>> Total 480+ OB's
>> Catalog Price $1.00 - 8 pages

<u>French Revolution Catalog</u>:
> 1792-1799
>> Total 450+ OB's
>> Catalog Price $1.00 - 5 pages

<u>Napoleonic Wars Catalog</u>
> 1800-1815
>> Includes Russo-Swedish 1808-9 War, Russo-Turkish Wars, Continental Europe and Peninsular Wars.
>> Total 1,400+ OB's
>> Catalog Price $1.50 - 15 pages

<u>Miscellaneous American Wars Catalog</u>:
> 1775-1864
>> American Revolution, War of 1812, Mexican-American War, Franco-Mexican War.
>> (Civil War has a separate catalog)
>> Total 400+ OB's
>> Catalog Price $1.00 - 5 pages

<u>American Civil War Catalog</u>
> 1861-1865
>> Total 1,300+ OB's
>> Catalog Price $1.50 - 11 pages

<u>Post Napoleonic Catalog</u>
> 1816-1899
>> Russo Turkish War 1829, French in Morée, Belgian Revolution, French in Algeria 1830, Polish Revolution 1832, Carlist War 1834, 1848 Revolutions, Roman Republic 1849, Crimea 1854, Italian Revolution 1859, Prusso-Danish War 1864, Prusso-Austrian War 1866, Franco-Prussian War 1870, Russo-Turkish War 1877
>> Total OB's - 200+
>> Catalog Price $0.50 - 3 pages

<u>British Colonial Catalog</u>
> 1861-1865
>> Northwest Indian Frontier, 2nd Afgan War, China, Boer, Zulu and others
>> Total 75+ OB's
>> Catalog Price $0.50 - 2 pages

<u>Naval Orders of Battle Catalog</u>
> 1688-1801 - 500+ OB's
> WWI   - 30 OB's
>> Catalog Price $1.00 - 8 pages

<u>World War I Catalog</u>
> 1905-6 Russo-Japanese War:
>> 10 OB's
> 1914-1918 World War I
>> Eastern & Western Front
>> Russian, Austrian, German, Serbian, Romanian, American, British & French
>> Total 200 OB's and growing
>> Catalog Price $0.50 - 5 pages

<u>Spanish Civil War</u>
> 1936-1939 125 OBs
>> Catalog Price $0.50 - 3 pages

<u>World War II Catalog</u>
> 1939-1945
>> Naval, Pacific, Mediterranean, North Africa, Europe.
>> Finnish, Russian Front, Vichy French, USMC in the Pacific, lots of TO&E material on every army.
>> 2,400+ OB's (over 8,500 pages of data)
>> Catalog Price $3.00 - 41 pages